Pro Tools 9

CW01597332

Pro Tools 9

The Mixer's Toolkit

Nathan Adam
Kevin Ward

ELSEVIER

AMSTERDAM • BOSTON • HEIDELBERG • LONDON • NEW YORK • OXFORD • PARIS
SAN DIEGO • SAN FRANCISCO • SINGAPORE • SYDNEY • TOKYO

Focal Press is an imprint of Elsevier

Focal Press

Focal Press is an imprint of Elsevier
225 Wyman Street, Waltham, MA 02451, USA
The Boulevard, Langford Lane, Kidlington, Oxford, OX5 1GB, UK

© 2011 Nathan Adam. Published by Elsevier, Inc. All rights reserved.

No part of this publication may be reproduced or transmitted in any form or by any means, electronic or mechanical, including photocopying, recording, or any information storage and retrieval system, without permission in writing from the publisher. Details on how to seek permission, further information about the Publisher's permissions policies and our arrangements with organizations such as the Copyright Clearance Center and the Copyright Licensing Agency, can be found at our website: *www.elsevier.com/permissions*.

This book and the individual contributions contained in it are protected under copyright by the Publisher (other than as may be noted herein).

Notices

Knowledge and best practice in this field are constantly changing. As new research and experience broaden our understanding, changes in research methods, professional practices, or medical treatment may become necessary.

Practitioners and researchers must always rely on their own experience and knowledge in evaluating and using any information, methods, compounds, or experiments described herein. In using such information or methods they should be mindful of their own safety and the safety of others, including parties for whom they have a professional responsibility.

To the fullest extent of the law, neither the Publisher nor the authors, contributors, or editors, assume any liability for any injury and/or damage to persons or property as a matter of products liability, negligence or otherwise, or from any use or operation of any methods, products, instructions, or ideas contained in the material herein.

Library of Congress Cataloging-in-Publication Data

Adam, Nathan.
Pro Tools 9: the mixer's toolkit/Nathan Adam and Kevin Ward.
p. cm.
ISBN 978-0-240-81870-2
1. Pro Tools. 2. Digital audio editors. I. Ward, Kevin, 1968– II. Title.
ML74.4.P76A32 2011
781.3'4536—dc22 2011004796

British Library Cataloguing-in-Publication Data

A catalogue record for this book is available from the British Library.

ISBN: 978-0-240-81870-2

For information on all Focal Press publications visit our website at *www.elsevierdirect.com*

Printed in the United States of America

11 12 13 14 15 5 4 3 2 1

Typeset by: diacriTech, Chennai, India

Working together to grow libraries in developing countries

www.elsevier.com | www.bookaid.org | www.sabre.org

ELSEVIER BOOK AID International Sabre Foundation

Contents

Acknowledgments

Thanks to God, my beautiful wife Jamie and my babies Araunah and Symphony for providing for and putting up with my late-night editing, recording, and producing project, after project, after project. I could not hope for a more supportive family.

Thanks to Kevin Ward for a willingness to teach your decades of experience to the audio students of the world, and being a constant inspiration.

Finally, an extra special thanks to Chris Haseleu and the rest of my students and colleagues at Belmont and MTSU for helping me hone and refine my techniques and teaching style over the last (and best) decade of my life. I have genuinely been blessed with the lottery of all job descriptions, thanks to you.

Nathan Adam is a professor of Audio Engineering at Belmont University. His work has extended to albums, tracks, TV and radio spots for artists including Collin Raye, Debbie Boone, Neal McCoy, the Grascals, Jim Stafford, Larry Gatlin, Dusty Rogers Jr., Moe Bandy, Carrie Tillis, Les Brown Jr., and many others. He is currently co-owner of the Walnut House recording and event complex in Murfreesboro, Tennessee. Visit www.nathaneadam.com for news on current recording projects.

Nathan Adam

First of all, I would like to thank my heavenly Father for being in control (although I constantly have to remind myself). Thank you Janna for being the love of my life, my constant companion, and tireless encourager. Thank you Kaelyn Sarah for being such a sweet surprise … I've never knew love quite like this, little princess.

I'd also love to thank my friend and fellow brainstormer Nathan "Adan" Adam … Seems every time we get together, we get excited and motivated about all of our ideas. I appreciate you, man. I'd like to extend a *big* "thank you" to my family … I love you all very much and wish I could see you more. Also, thank you to all of my unknowing mentors. I'm watching you closely. Thank you for living an encouraging life. Lastly, thank you to all of my www.MixCoach.com subscribers and readers. Your feedback keeps me going.

Kevin Ward has been sought out for his creative mixing skills from the likes of Dolly Parton, Vince Gill, Willie Nelson, Judy Collins, Richie McDonald, the Prague Philharmonic, and countless others since he started professionally in 1991. He is renowned in Christian music for his work with the Christ Tabernacle Choir, Calvin Hunt, Janet Paschal, and LordSong, and the Gospel Music Association has given him its highest honor, the Dove Award as a Producer. He has been nominated for his songwriting as well. He is also the founder of MixCoach.com and co-founder of Vine Records.

For more information, visit www.ProTools9MixersToolkit.com

Kevin Ward

In This Chapter

1

Session Overview

Intro and Overview

Welcome to *Pro Tools 9: The Mixer's Toolkit*. Four years ago, we looked at the advanced production techniques used in the creation of modern records in *Multi-Platinum Pro Tools*. Since that time, the entire audio world has advanced with a number of tools, and as a result skills, so as to make these techniques available to the masses.

From the "I Am T-Pain" Autotune iPhone app to the ability of the free GarageBand software to automatically time align your tracks, it's never been a more exciting (surreal?) time to be an editor in the recording industry.

On the flip side, this also means that your little brother now has the ability to "mix" a track. It means that every engineer, in every bedroom, has the software to pocket a drum kit, guitars, and even those pesky pianos.

However, what every bedroom engineer does *not* have is a firm understanding of how and when to use these tools and tricks in a mixing situation. What we're going to teach you in this book (and accompanying Web site, videos, and demo sessions) is an understanding of the tools available in Pro Tools 9 to create a professional editing and mixing workflow. Even more importantly, you'll be developing your ear to be able to hear the subtle differences between a professionally mixed and edited track and one that has just been slapped together by the armchair engineer. So, in this book, we are going to take you through a full song mix, utilizing all of the tricks at our disposal in making a rocking, radio ready mix.

From the old standbys of manual pocketing and Beat Detective to the new tools of Elastic Audio, we will bring you up to speed on the killer tools used by the professional Pro Tools engineer. Specifically, we're going to take you through the entire mix process, including track arrangement, equalization, compression, automation, and so much more! We are going to explore vocal mixing, parallel drum compression, advanced bussing techniques, and home studio pre-Mastering.

© 2011 Nathan Adam. Published by Elsevier, Inc. All rights reserved.

Demo Files and Web Site

We know that people have different learning styles, so we wanted to make sure you get the best of all worlds when working through *Pro Tools 9: The Mixer's Toolkit*. We started where you are too, and we know that some processes (such as the exact timing of automation moves in a vocal ride) are hard to grasp in the nontiming-based nature of text. As a result, as the first purchaser of this book, you get an included one-month membership to www.mixcoach.com, where you will have access to hours and hours of streaming video examples and tutorials covering these and other advanced mixing and production techniques. You'll also be able to access all of the audio files and other sample sessions to further enhance your knowledge. Audio technology and techniques never stay static, so come, visit us at www.mixcoach.com to truly enhance your learning experience.

Who You Are?

In order to write any book, assumptions have to be made about the kind of person who's going to read it. We've written this book to speak directly to the types of new audio production students we encounter every day at audio production programs like Belmont University (www.belmont.edu) and Middle Tennessee State (mtsu.edu/~record). We believe you are an aspiring audio producer or engineer with your own DAW, and you've started recording projects ranging from guitar or vocal demos to maybe even some full bands. You want your finished mixes to sound on par with the commercially released tracks you hear on the radio every day, and you work to improve your skills with the aim of engineering for a living one day.

So, in this book, we're going to operate under the assumption that you have the ability to launch sessions, operate the transport (play, fast forward, rewind, etc.), and even use the various modes and basic tools of Pro Tools (or whatever DAW you're going to follow along with).

What if you're a total newbie though? Well, we have some good news for you too. While this book is *not* written to the "100% new to digital audio workstations like Pro Tools" crowd, with the included membership to www.mixcoach.com, you *do* get access to a wealth of "Introduction to Pro Tools" videos that will bring you up to speed in no time flat. After you spend a few hours working through those videos and their included sessions, you'll be ready to tackle the more advanced editing and mixing tools we cover in this book.

Shortcuts and Menus

One of the biggest things we've noticed when working with new interns at either of our studios (www.walnuthousemusic.com and www.kevinwardmusic.com) is how *painfully slow* they tend to be at editing and mixing in a DAW. Even graduates from top audio schools have become accustomed to spending long hours of leisurely paced time in the studio producing 1–2 tracks over an entire semester. In the real world, this is a great way to make a *bad* impression in a professional recording environment. As a result, we're going to drill the need for you to add the skill of *keyboard shortcuts* to your personal mixing toolkit.

As a majority of the professional recording engineers work on the Mac version of Pro Tools for its enhanced stability and reliability, we will first reference the Mac shortcuts, with the Windows shortcuts shown in parentheses for those of you working on that platform. For those of you new to extensive keyboard shortcut references, this won't take long to learn, but it must be practiced every time you open Pro Tools. After a day or so though, they will become second nature, and you will be well on your way to impressing the next professional producer you work with.

First, you can and absolutely *should* attempt to learn any shortcut that will reduce the number of menus you have to click on during a session. The technique I teach to students and interns goes something like this.

When you find yourself clicking on a menu to perform any Pro Tools function (like Separate Region, Save Session etc.), take note of the symbols to the right of the function.

Figure 1.1 The Shortcuts Are Listed to the Right of the Pro Tools Function.

These show you the keyboard shortcuts that can be performed, rather than clicking on the menu every time. Then, if you use the function more than once in a session, find it in the menu, then rather than clicking it, force yourself to use the keyboard shortcut. Undo, then use the shortcut again to start embedding it in your mind. If you follow this process, within a session or two, you will be learning shortcuts at a rapid pace and dramatically enhancing your mixing efficiency.

Use the reference below for the symbols and shortcuts on Macs and Windows. Mac keyboard shortcut symbols are shown in Figs 1.2–1.5.

Figure 1.2 Control or Ctrl Key.

Figure 1.3 Alt or Option Key.

Figure 1.4 Command or Cmd Key.

Figure 1.5 Shift Key.

Here's a few more tips that will make learning shortcuts even easier.

1. Always use the "Alt" term for both platforms. It does the same things on both platforms, and all new Macs also label the key with an Alt name, as well as the classic Option name. It's also easier to remember that the "Alt" key makes many PT functions (Solo, Mute, etc) effect "ALL" tracks.
2. When you see a shortcut using the Mac Command button, you can almost always assume it will be with the Control button on Windows. For example, Copy on the Mac (Cmd+C) becomes Ctrl+C on Windows.
3. When Pro Tools (rarely) uses the Mac Control button, it always translates to the "Start" or "Windows" button on most PCs. So, a shortcut such as Ctrl+Shift+T on the Mac for Keyboard Lock will become Start+Shift+T on the Windows version of Pro Tools.

Figure 1.6 Ctrl+Shift+T on Mac, Start+Shift+T on Windows.

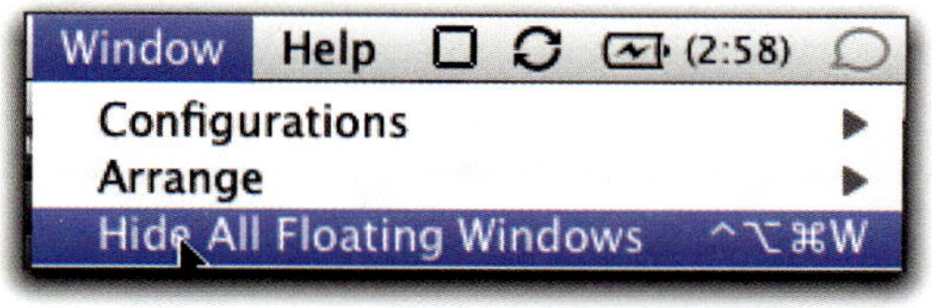

Figure 1.7
Ctrl+Alt+Cmd+W.

Finally, we will be abbreviating references to menus and shortcuts in the following manner. Referencing the image above, we would tell you to go to *Window Menu>Hide All Floating Windows.* So that you should get ready to launch into the session. Open up the included tracks, and let's get mixing!

In This Chapter

2

Session Setup

Listening Through the Original Tracks

Depending on who you are talking to, the 1970s were the greatest decade in recent history. Oh sure, we had to put up with all the crappy disco music we would ever need, but for a few blissful years, little programmable sequencers freed us from plague of human drummers. (Ba-Zing!). But seriously, as a wannabe drummer myself, I sympathize with the plight of any musician who has to have four limbs playing different rhythms simultaneously. Above all, however, I sympathize with the engineer who has to fix a challenging drum performance. Fortunately, in this song, we have a good rock drummer, who just needs to be tightened up. Of course, that's why you're here.

With that, fire up the included Pro Tools session, and you should see the tracks laid out ready for your masterful touch (Figure 2.1).

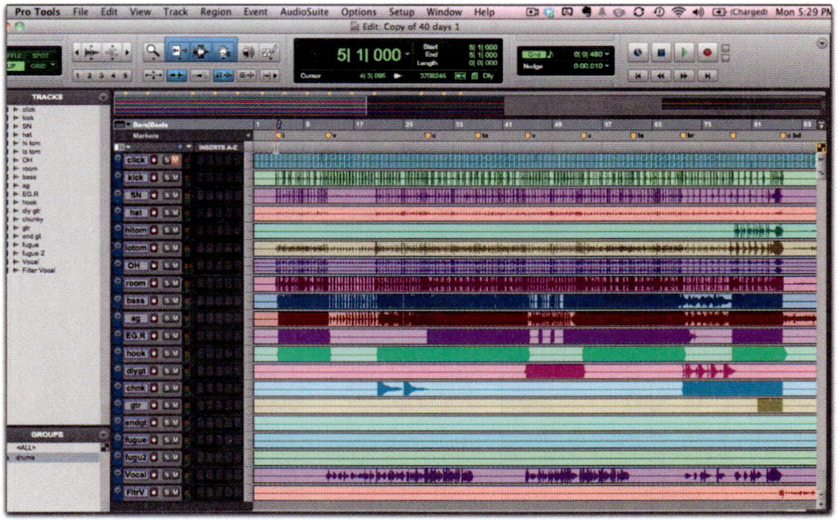

Figure 2.1 Your Session Should Start Out Looking Like This.

© 2011 Nathan Adam. Published by Elsevier, Inc. All rights reserved.

Now as we've already stated, this is *The Mixer's Toolkit*, not "Intro to ProTools for Toddlers." So, we are not going to walk through how to arrange tracks by clicking and dragging on their name, or in the Tracks window, but it is worth noting the order that we chose to arrange the tracks in. This is a fairly standard track arrangement, with the click track on top followed by the drums, the bass, acoustic and electric guitars, and finally vocals. It is a pretty standard track layout for pocketing, as our reference track (in this case, the click) resides on top, ready for us to edit the drums against.

Everybody wants to jump right into the advanced editing and mixing techniques as soon as they sit down at their first digital audio workstation. However, as we saw last time, if you don't set your software up properly and in a way that is conducive to the work you're going to do, the technology can often frustrate your creativity. There is nothing more maddening than pressing a flurry of keyboard shortcuts expecting voodoo to happen as notes and tracks magically align, only to have your screen rearrange, audio disappear, and your computer freeze up, all because you didn't take just a few moments to optimize your set up. Lazy, lazy, lazy.

Don't feel bad though because we are about to kick things in gear. First things first, go ahead and play through the track, so you can get a feel for the instruments, vocals, and frequency ranges that everything is occupying. Your session should look something like this (see Figure 2.2).

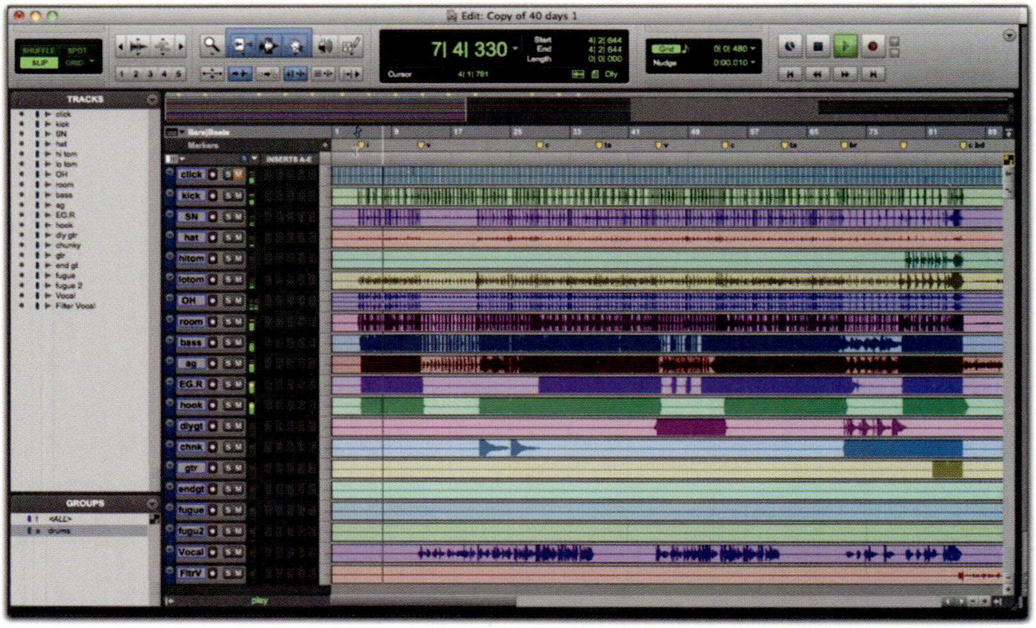

Figure 2.2 It All Starts with a Great Song ... In a Garage.

The Alt and Shift Key Secret

The beauty of using a session like this to learn about mixing is that it has many great elements, as well as elements that need some sweet mixing action. Seriously, like the drums.

In order to really focus on the drums, let's select all of the nondrum tracks and mute them, of course you could simply click on the individual mute buttons, but that would be slow and painful. This brings us to our first Pro Tools mixer's shortcut that you will use hundreds of times in a given session. First let's select the range of instruments between the bass guitar and the filter vocal. Click on the name of the bass guitar track, then hold down your Shift key while clicking the filter vocal track at the bottom (Figure 2.3).

Figure 2.3
(a, b) The Magic of the Shift Key.

Voilà! Holding the Shift key selects everything between your two selections. The beauty of this is that it works in most every Pro Tools editing and mixing situation you would desire. For example, want to select all audio between two markers? Simply select the first marker by right-clicking on it, then Shift+click the later marker; all audio between them is automatically selected.

Back to muting our nondrum tracks, we get to our second key shortcut that utilizes the Alt and Shift keys. If you've been using Pro Tools for any amount of time, you know that holding down the Alt key (on either Mac or Windows) will affect all tracks in the session. What many people don't know is that when you add the Shift key as well, your actions will affect all *selected* tracks.

We use this to our advantage now by holding the Alt and Shift keys while clicking the mute button on any of the selected tracks. In only a few short clicks, we have muted all of the nondrum tracks (Figure 2.4).

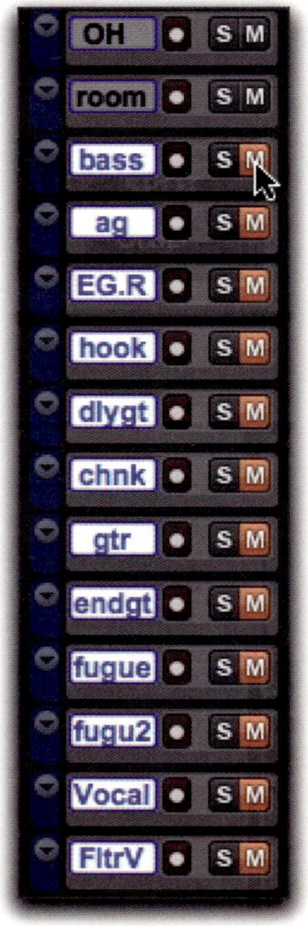

Figure 2.4 One Click Beats a Dozen Clicks Any Day.

The last step in preparing to edit our drums is to hide all of these unused tracks and fill the screen with only the drum tracks. Because you are all aspiring to be professional, efficient mixers let's do this with two strokes (as opposed to the dozen or more you were about to use ...).

Still holding down the Alt and Shift keys, click on the little circle to the left of any of your unused tracks in the Tracks window. You will see them instantly disappear from your Edit window (Figure 2.5).

(a)
(b)

Figure 2.5 (a, b) Quickly Hiding a Large Group of Tracks.

That's at least a minute of clicking you've shaved off with Alt+Shift already, huh?

The Fat Finger Secret

Next, vertically resize all of the remaining tracks with our next secret ninja shortcut … I call it the *fat finger*.

Fat finger is a short way of saying to press the Control, Alt, and Command (Win: Ctrl+Alt+Start) buttons all at once, as though your finger was too fat to press just one of them. Get it? This is one of the least-known shortcuts in the entire Pro Tools universe, but it does so much in so many different situations. For example, right now you can hold down *fat finger (Mac: Ctrl+Alt+Cmd, Win: Ctrl+Alt+Start)* and hit your down-arrow key on the keyboard to instantly resize all remaining tracks to fill the vertical space of the Edit window (Figure 2.6).

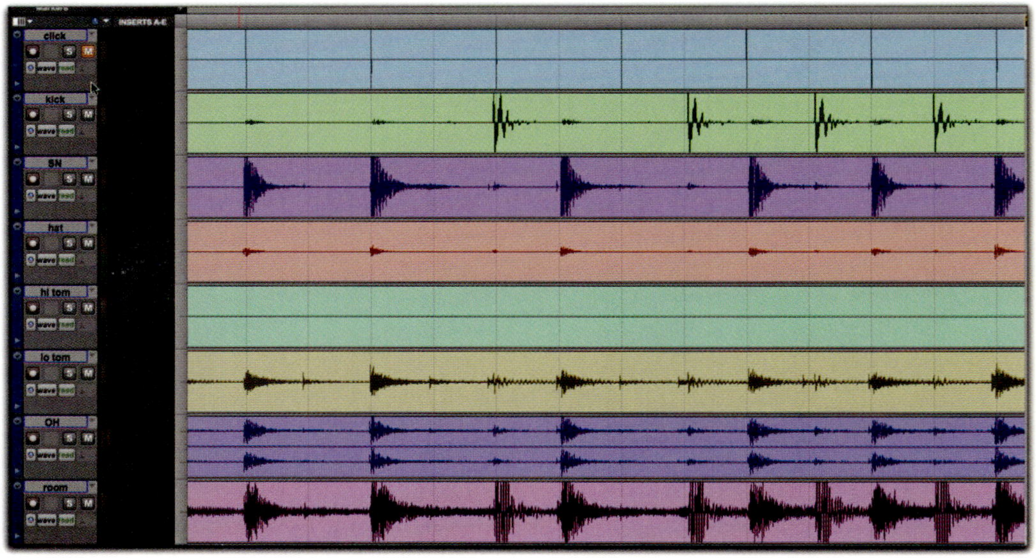

Figure 2.6 That's Just the First of the Fat Finger Tricks.

Drum Editing Prep

When you are an editor, you will always be hiding and showing different combinations of tracks so that you can focus your attention on just the current edits. You will also resize tracks with great regularity both vertically and horizontally. As a result, this fat finger trick is just one of the speed techniques, which separates the men from the boys.

Now that your drums are up and ready, let's create a drum group so that any edits we make to one drum track is made to all the drums simultaneously. If we were to try to edit the drums individually, all of the delicate phase relationships achieved with these fantastic garage miking techniques would be lost. Using the Shift+Click method we discussed earlier, select all of the drums and hit Command G (Win: Ctrl+G). Select all of the drums, name your group "Drums," link the Edit and Mix windows, and click OK. Once you've done that, you should see it in the Groups window, and any selections or edits made to one drum track will now affect all of them (Figure 2.7).

(a)

(b)

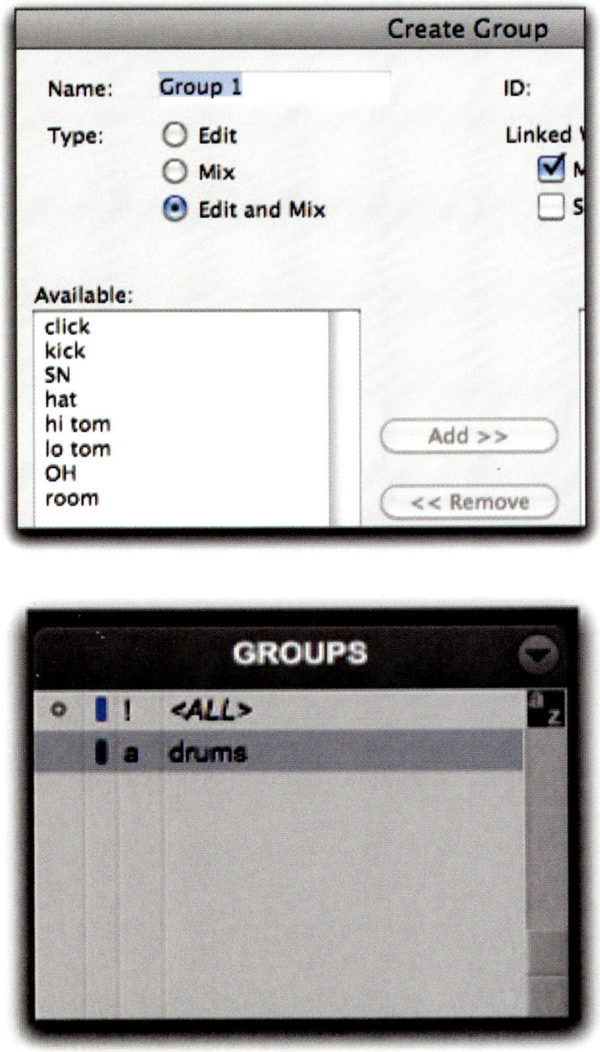

Figure 2.7
(a, b) Groups Are a Mixer's Best Friend.

Creating an Escape Route

First things first, before we go editing this track all to death, we need to make ourselves an escape route. That's to say, often times you will find yourself mixing a track that you did not have the pleasure of recording. As a result, it is very normal to polish the track to be the best that it can possibly be. However, what happens if the drummer or guitarist or vocalist comes back and says "I meant to play it like that"? Well, if you haven't created a backup of the unaffected session, or a playlist of the original tracks, then there's simply no easy route to get them back.

Obviously, the simplest solution is to turn on the Auto Backup preference found under the Setup menu > Operation tab (see Figure 2.8).

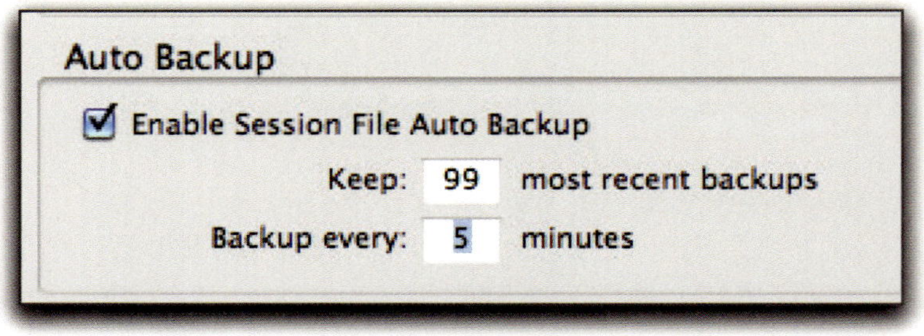

Figure 2.8 Just Go Turn It on Right Now. Don't Wait. Check for It Every Time You Open Pro Tools, or Regret It Later.

This preference will create a Backups folder in your main session folder, and store a sweet, sweet safety copy every few minutes. There is no reason that I can think of to not turn this on.

However, it's not necessarily the best way to keep alternate versions of your tracks. For example, who wants to go back through hundreds of old sessions looking for original drum tracks? Certainly, not me.

So instead, we'll use an often overlooked feature of Pro Tools called playlists.

Play listing is a function that allows you to create and access alternate takes of a performance. Simply click on the down arrow to the right of the track name and you will see options for "New, Duplicate, and Delete Unused."

While there is certainly a lot to be explored when it comes to using playlists for functions such as solos and vocal comping, for now we are just going to make a duplicate of our existing drum tracks (Figures 2.9 and 2.10).

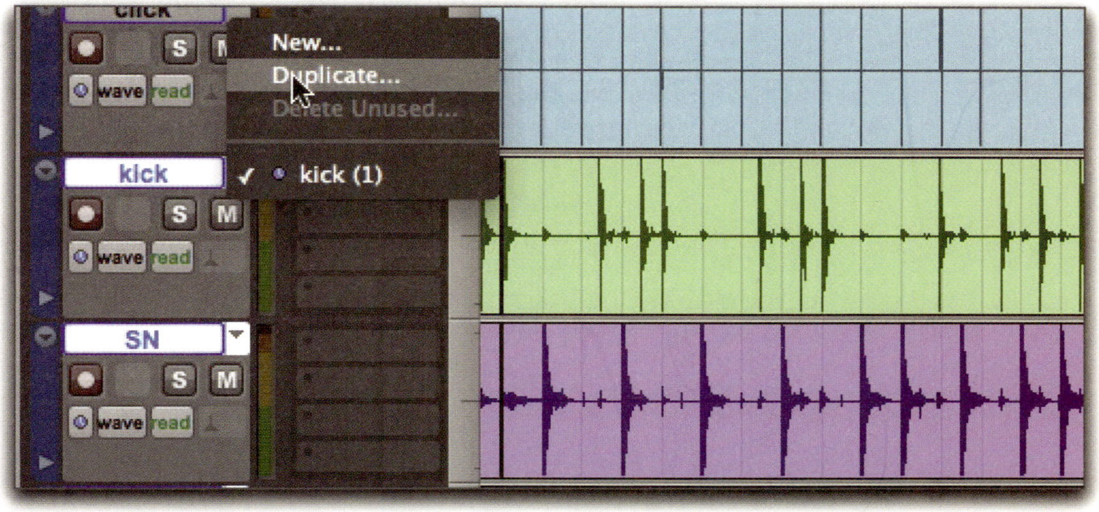

Figure 2.9 Duplicating a Playlist Is a Quick Way to Move Forward with Any Dangerous Edits.

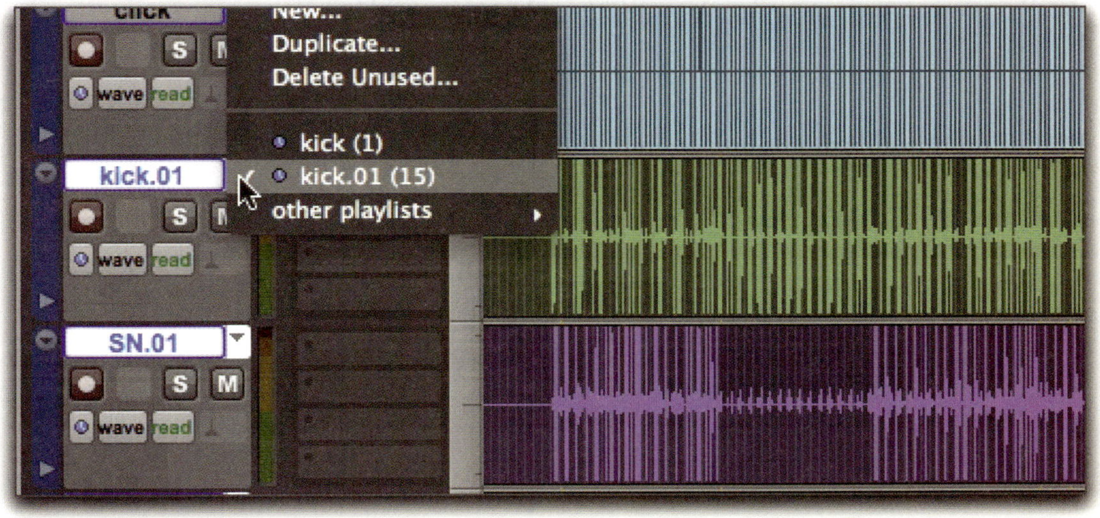

Figure 2.10 Secure in the Knowledge That the Originals Are Hiding Behind the Scenes.

Once you have duplicated your playlists, you will see that your tracks now have a .01 beside them. You can rename them if you like so that these playlists are labeled with the words "Edit" or "Original," etc. Just be sure to use a naming convention that makes sense to you or follows the NARAS guidelines for naming tracks. There is nothing worse than trying to find an old take on an alternate playlist with an ambiguous name like "Audio_07.3."

The beauty of duplicating this drum playlist is that we can access the originals at any time with just a click. So, if we get halfway through pocketing these drums and find we have accidentally pushed the entire chorus out of sync, we can simply switch back to the original playlist, copy the chorus, and switch back to our edit list to paste it in. Crisis averted.

In This Chapter

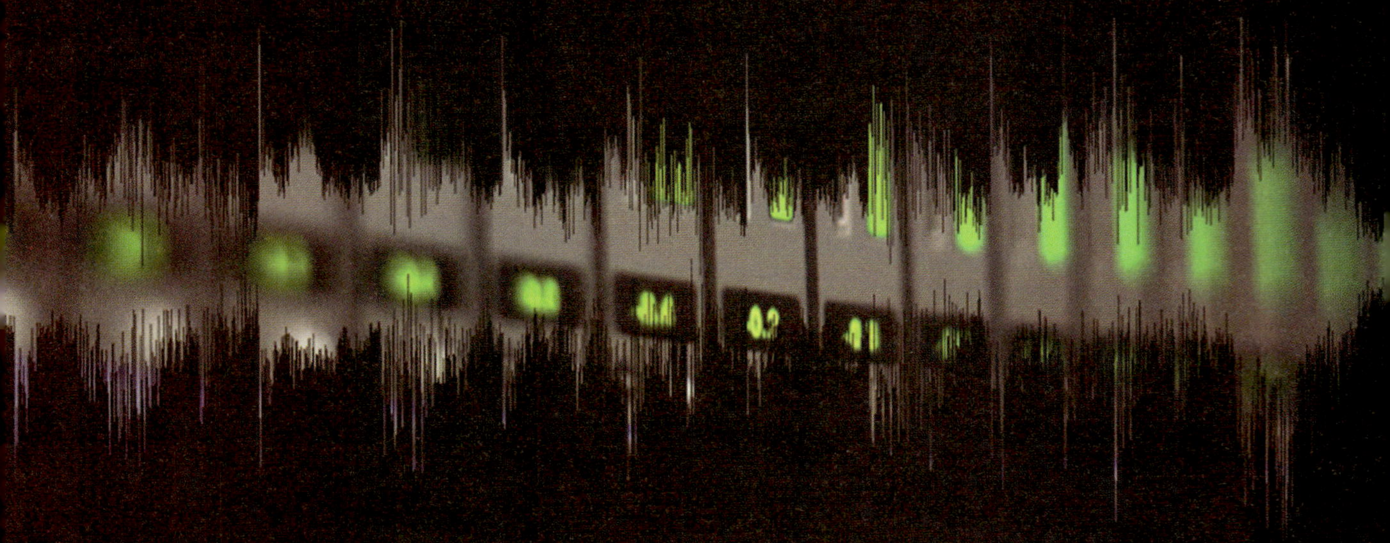

3

Drum Editing

Manual Drum Pocketing

To start editing the drums, let's zoom in on this drum performance and compare the visual of the transients against the grid tempo that we recorded the song to. It's pretty plain to the eye (if not the ear) that this drummer rendered a fine performance … for someone who was in a hurry. As is common with many rock drummers, he has a tendency to stay ahead of the beat by a significant margin.

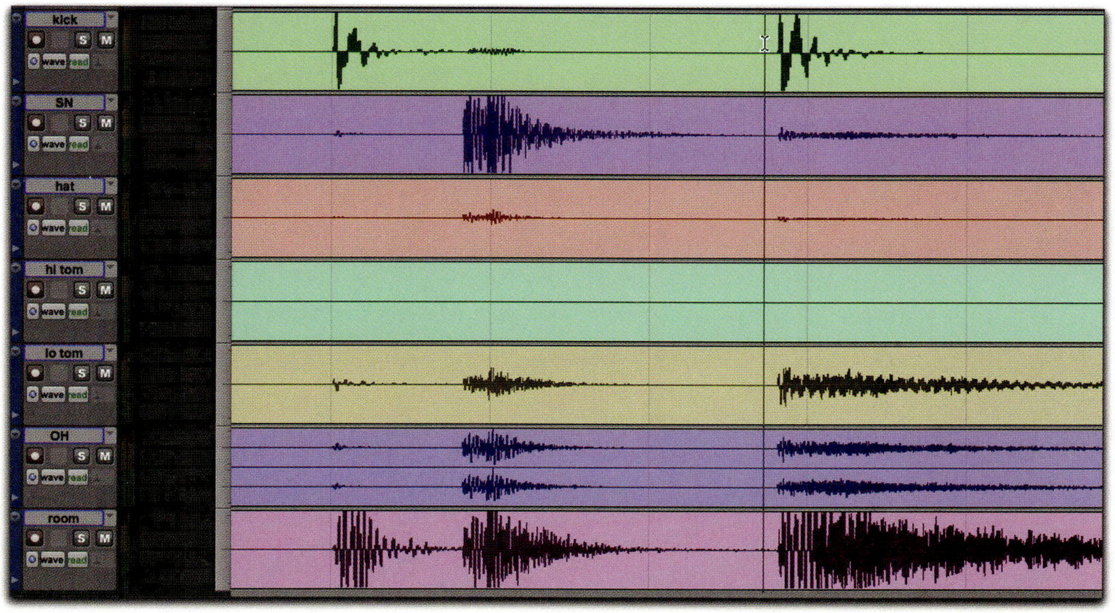

Figure 3.1 Whoa There Fella. Where's the Fire?

© 2011 Nathan Adam. Published by Elsevier, Inc. All rights reserved.

As a result, if we were to use a single editing technique (such as Beat Detective), we would find that many of the notes would get quantized to the wrong eighth note. This is a common scenario for the modern mixer, so we will correct this performance using a variety of drum editing techniques, including cutting, nudging, Beat Detective, and Elastic Audio.

To get started, we are going to work through the song section by section, identifying the problems and using the best tool for the job.

Starting with bar 5, we see that our kick and snare drum is almost 1/16 note ahead. That's a valuable piece of information but also a caution for you "jump the gun" types. See, we need to always use our eyes and ears when editing because if he consistently plays the snare drum a sixteenth note ahead and we attempt to autoquantize him, the software will tend to move him to the early sixteenth note, which would just make more work for us later (Figure 3.2).

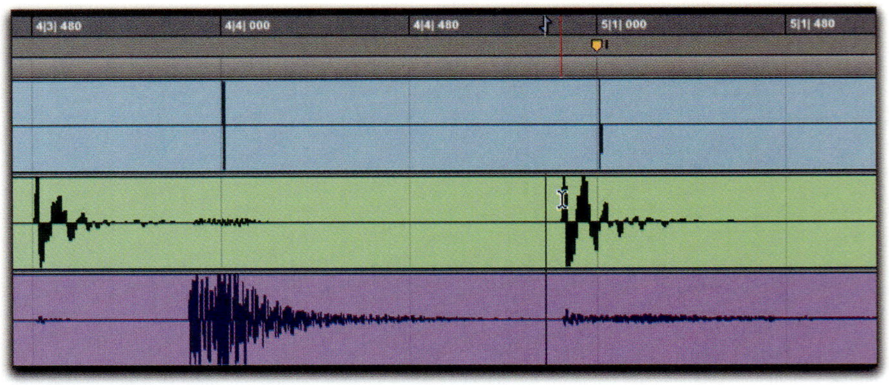

Figure 3.2 Not Pictured: John Bonham.

So, let's unmute the click track and listen down to the first eight bars or so of the drums. Then, adjust your click volume if necessary so that you are just feeling it in the background but able to hear any deviations that the drummer makes.

Well, that didn't take long. Right off the bat, the snare at bar four beat four came in a little early. So, let's quickly demonstrate a manual snare pocket using Tab to Transients.

Roughing in the Drums

Tab to Transients is a function in Pro Tools that attempts to find the next transient in the selected track every time you hit the Tab key. It is especially effective and useful on drum tracks that have very clearly defined transients. It's less reliable on sustaining instruments like pads, pianos, and strings, as their transients are often very slow and hard to detect.

However, for this snare drum, click your Selection tool in the snare region and you should see the selection line across all of the grouped drum tracks. Now, with the Tab to Transients button engaged, hit your Tab key once and you should see the timeline selector move to the beginning of your snare hit, ready to be chopped and moved.

(a)

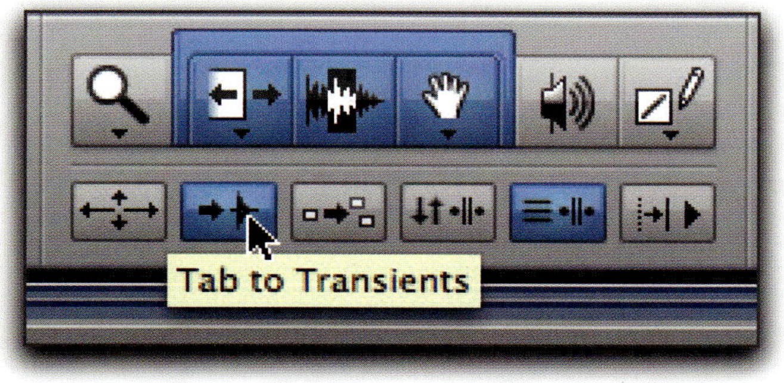

(b)

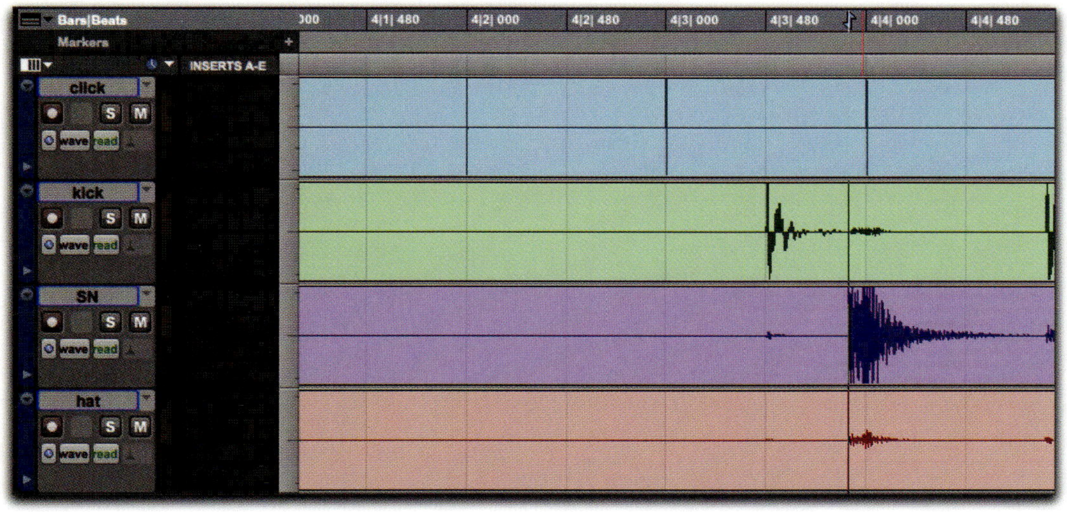

Figure 3.3 (a, b) Engage Tab to Transients, Then Chop Out the Early Snare.

Once it's at the transient of the snare, select Separate the Region at Selection, or as it's known among the pros … Command+E (Win: Ctrl+E). Quickly tab again to the next transient and hit Command+E again. Now, your snare region is ready to be moved into the proper time.

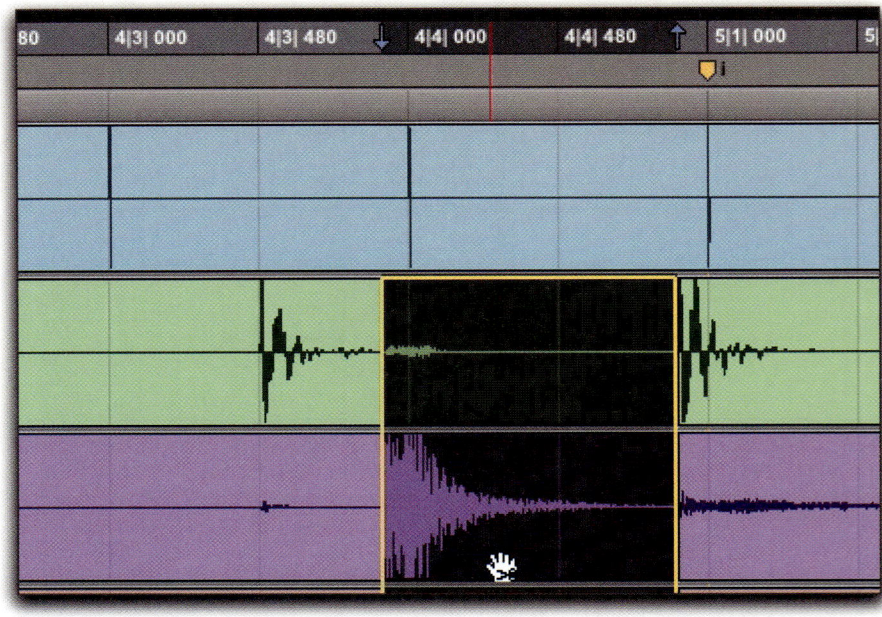

Figure 3.4 Move the Snare to the Correct Beat.

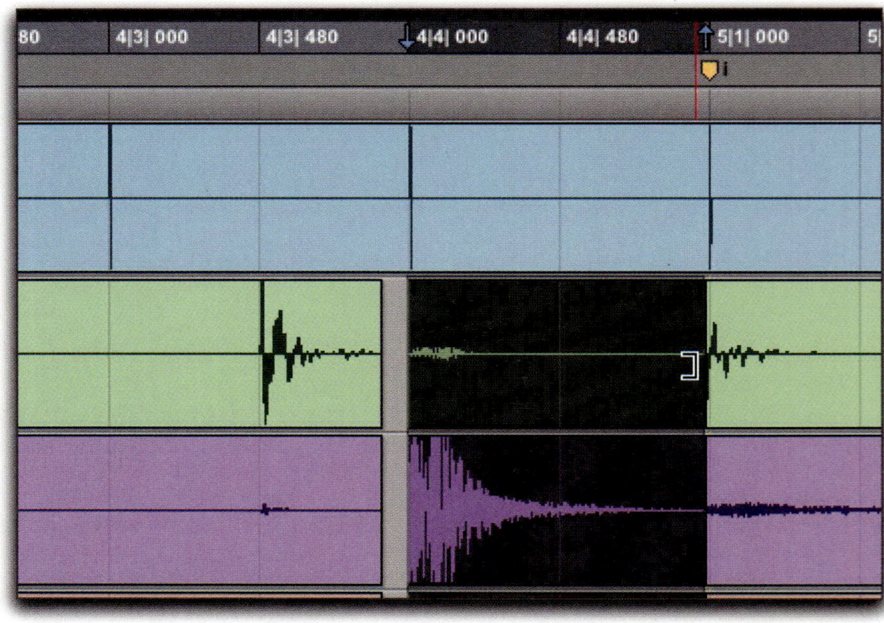

Figure 3.5 Then, Use the Trimmer to Reveal the Kick You Just Covered Up.

Now, of course, this has caused a second problem: we have hidden the kick drum that was a sixteenth note early. Ah well, an editor's work is never done. Since we said at the beginning of this book that you need to have a basic understanding of editing and Pro Tools, you probably already know to have your Smart tool selected for maximum efficiency in switching between the Trim, Selection, Grabber, and Fade tools.

Figure 3.6 Don't Be Dumb. Use the Smart Tool. It Says So Right on the Software.

Since you obviously know that, just hover over the left edge of the hidden kick drum and use the Trim tool to reshow it. Finally, trim the left edge of the snare back to fill in the gap.

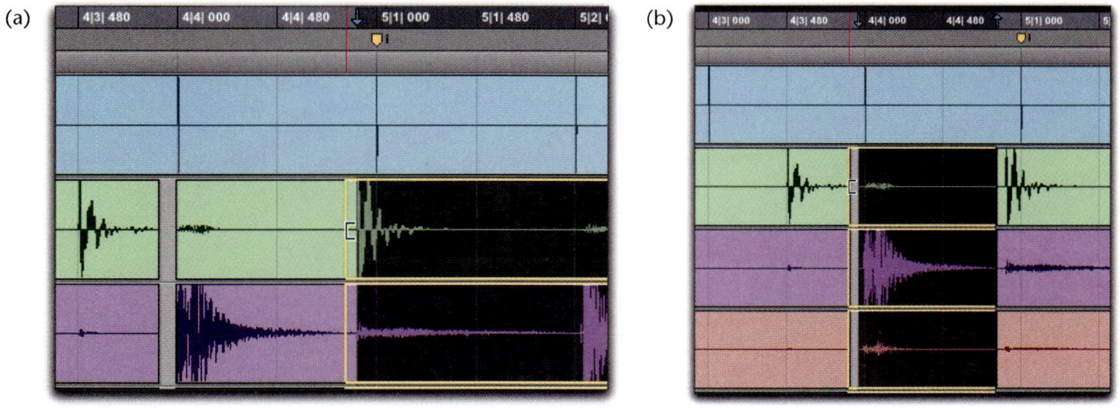

Figure 3.7 (a, b) Ahhh! There It Is! A Neatly Pocketed Snare Drum.

While you are at it, hover your mouse on the lower half of any of the regions until it turns into a Crossfade tool and create a small fade to hide the edit.

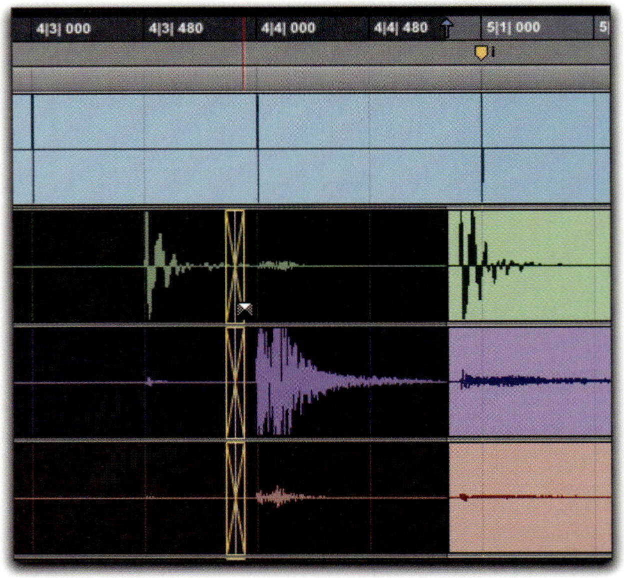

Figure 3.8
Hiding Edits with the Trim and Crossfade Tools.

So, since you now know how to manually drag a note into a place and hide the edit after the fact, let's use a combination of shortcuts to let Pro Tools do the heavy lifting for us on the next note.

Put your selector in the Kick track and use Tab to Transients to move to the next kick drum that is arriving a little early. Now, remember the Shift key? Hold it down and hit the Tab key again. As if by magic, Pro Tools has selected from the beginning of the early kick drum all the way to the transient of next hit on the Kick track.

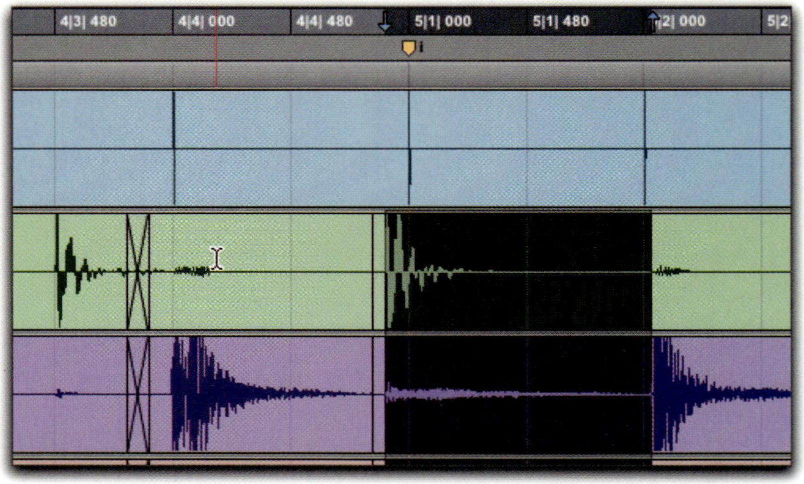

Figure 3.9
Shift+Tab Is a Quick Way to Grab Drum Beats.

Once again, separate the region with Command+E, and now we will use the quantize shortcut to move the selected regions to the nearest Grid value.

The Grid value is selected on your transport bar and can be set to anything from minutes and seconds to bars, beats, or even samples.

Since we are trying to move this kick drum to the nearest quarter note, make sure you have a Grid value of eighth notes assigned.

(a)

(b)

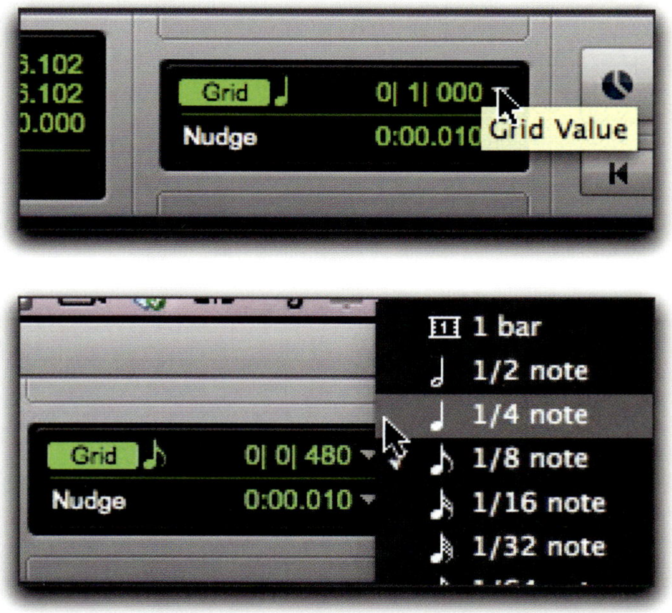

Figure 3.10
(a, b) Setting Your Grid Correctly Will Make Your Quantizing Fly.

Now, with the beat selected, it's a simple matter of hitting your quantize shortcut (Alt+0) to move the selected regions to the nearest quarter note.

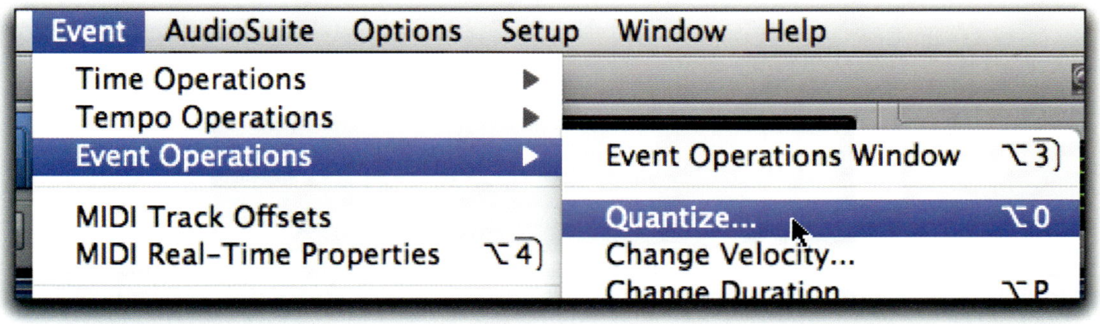

Figure 3.11 Quantizing.

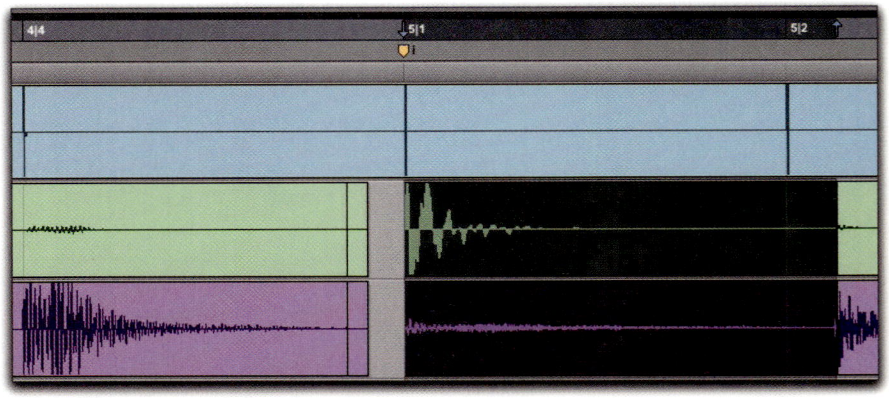

Figure 3.12 The Quantize Shortcut Will Shave Years Off Your Editing Life.

Now, of course, we've hidden the later note again, so we'll want to make sure we're in Slip Mode so that we can freely drag our region edges and drag the hidden snare drum back into view.

Figure 3.13 Stay In Slip Mode for 90% of Your Editing. Grid Is Useful for Building Loops and Arrangements, but Slip Allows the Most Flexibility When Editing.

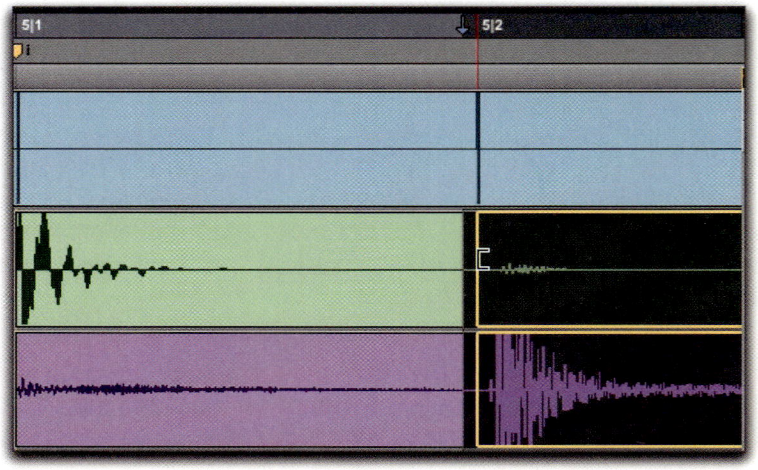

Figure 3.14 Theeeeerre It Is.

Drop a quick crossfade across both sides for now, but in a minute, we'll explore saving vast amounts of time with batch crossfades.

Now, as we move on over the next few bars, the drummer starts to play more consistently *behind* the beat, as opposed to ahead. Since he's fairly consistent in the amount he's playing behind, let's use our Nudge function to move the whole track forward a bit so that Beat Detective and Elastic Audio will have less work to do when we employ them shortly.

Select inside any of the drum tracks just before bar 7|2 and hold Shift while dragging left to around bar 5|3. All of the drum tracks will be selected, ready for separating and nudging.

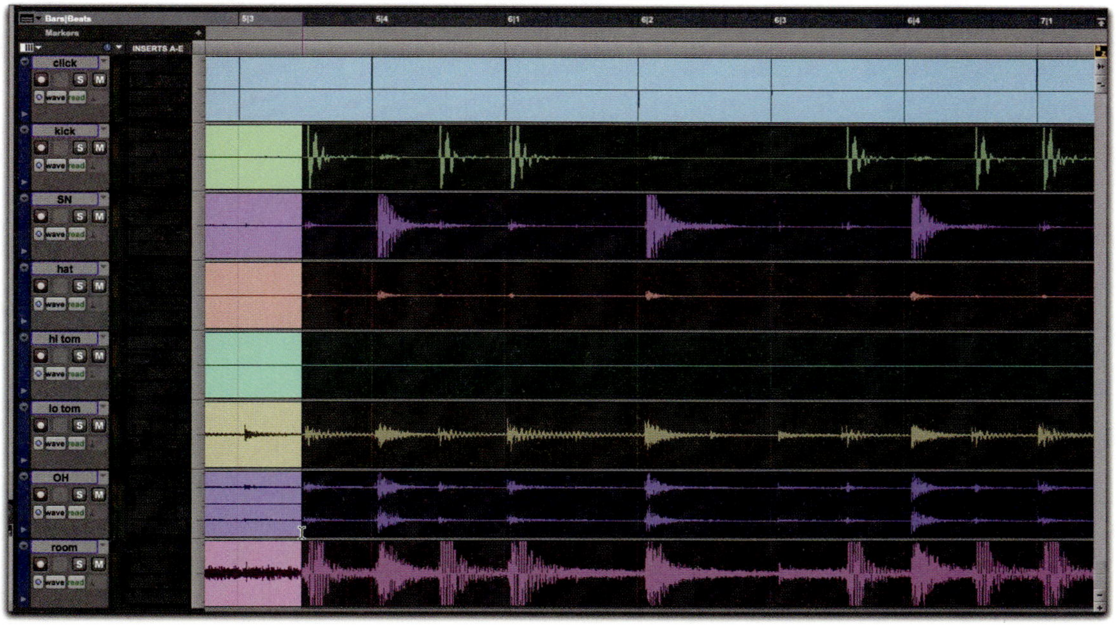

Figure 3.15 The Playing Behind the Beat Section Needs a Bit of a Nudge.

Now let's set our nudge value. I like to set my nudge value to around 10 ms when editing drums. I find it to be a small enough increment to create subtle moves or audible shifts in just a few clicks of the plus (+) and minus (–) keys on the keyboard.

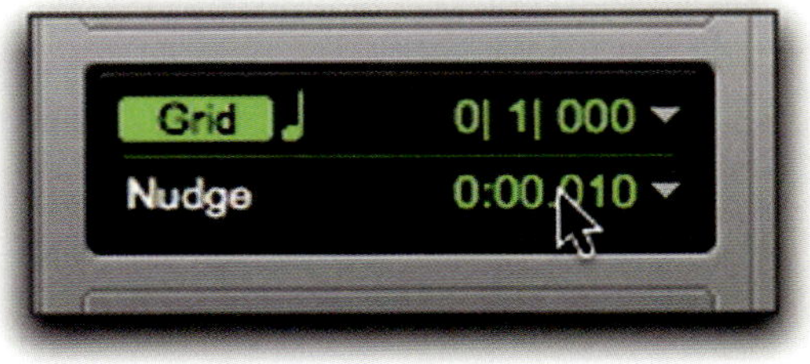

Figure 3.16 You Can Nudge by Virtually Any Amount, but 10 ms Is a Solid Choice When Editing Live Instruments That Don't Adhere to the Tempo Perfectly.

Separate the entire selection with Command+E (Win: Ctrl+E) again and use the plus and minus keys to nudge the audio until the next snare hit occurs directly on the grid line of 5|4.

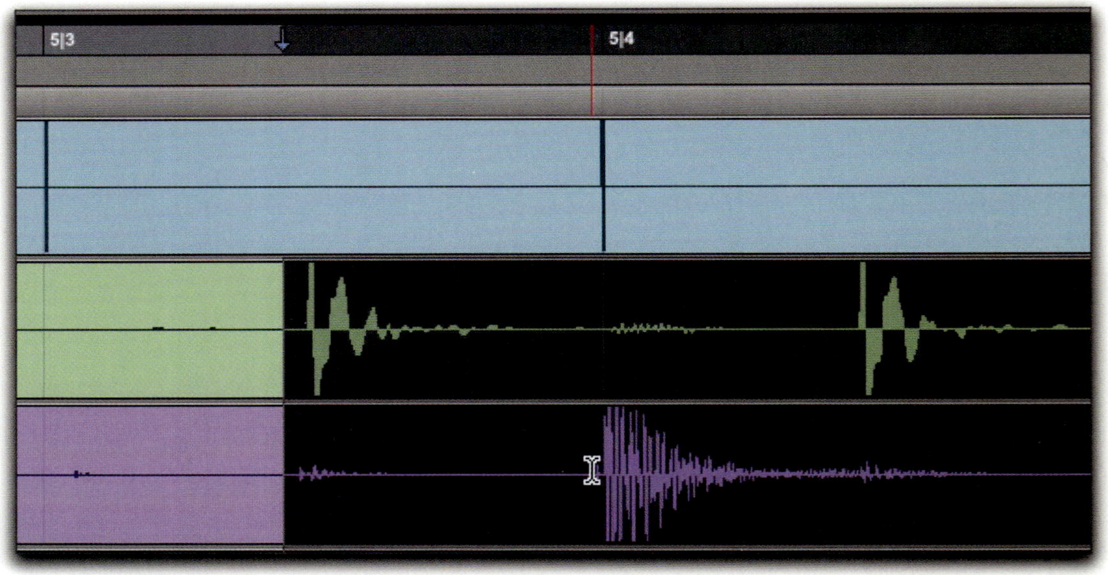

Figure 3.17 It's That Easy to Pocket Some of These Tracks.

In terms of manually nudging an errant drum beat into place, it's really as simple as that. Of course, in my other book, *Multi-Platinum Pro Tools*, I showed you just how in depth we take this type of manual pocketing on a major label track. From literally separating out every individual beat to manually pocketing (aligning) each one to within 8–20 ms of the click track, it is normal to spend several hours manually pocketing not just the drums, but all the other instruments and vocals as well.

If your project budget dictates the time and effort to invest the 90% of extra time for the additional 10% of good "feel" you can get from a quality manual pocketing, then by all means, go for it. Most modern projects are on tight budgets and timelines though, so let's explore a few of the amazing new tools that Avid and other DAWs have recently given us to dramatically speed up the audio timing correction process.

Fixing Timing Trends

Moving forward, you'll see that he's starting to fall more on the beat, with enough variation that we know he's a normal rock drummer. As we scroll through the song though, our objective now is to identify and correct general *trends* where he starts to get regularly ahead or behind the beat.

When we find a section where this is occurring, it's now a simple matter to do a quick Shift+Drag selection over the offending drum tracks and nudge them a few milliseconds forward or backward to bring his trend into time. Then, once we have quickly nudged the trends closer to the correct timing of the Pro Tools grid, we'll be able to swiftly engage the services of Beat Detective or Elastic Audio to make the track rock solid.

For example, between bars 14|2 and 15|4, he starts to trend late again. Now that we've learned the routine, simply select (Shift+Drag), separate (Mac: Cmd+E, Win: Ctrl+E), and nudge (+ & – keys) it forward a few clicks until they're closer to in time.

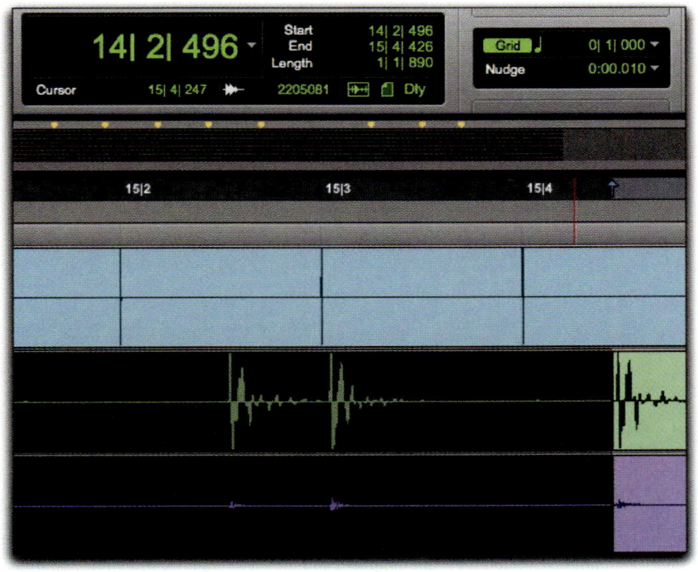

Figure 3.18 The Start and End Counters to the Right of the Main Counter Show Exact Start and End Times of the Current Selection. Use Them to Follow Along.

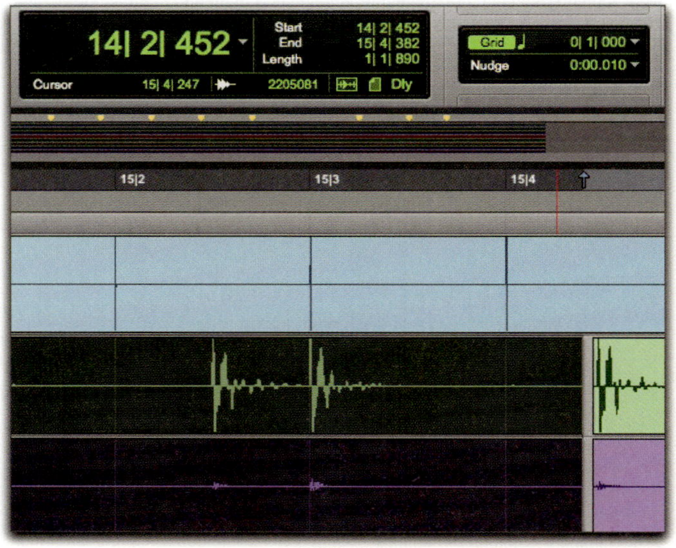

Figure 3.19 Two Nudges at 10 ms Moves the Kick Drum Right into Place.

Once it's in place, hide the edit with a crossfade and move forward in the track, identifying and correcting trends.

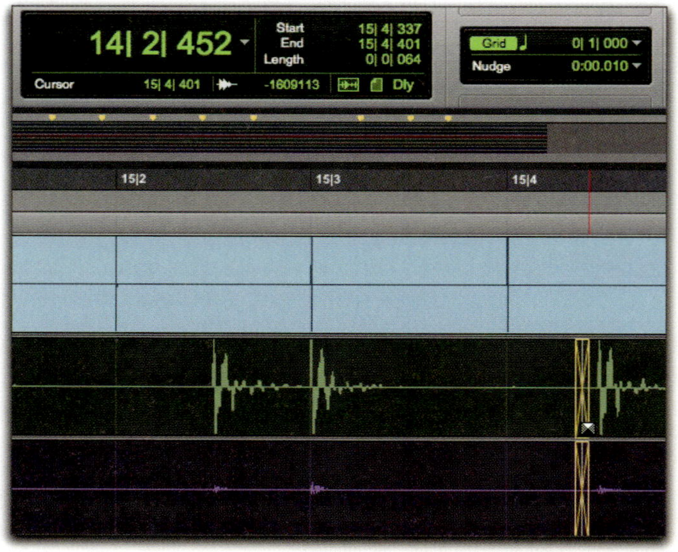

Figure 3.20 Always Clean Up After Yourself.

Remember, the nudge pass we're making now to get the drums generally closer to the beat will make our next pocketing pass go significantly smoother. That said, let's look at a few more shortcuts from the Mixer's Toolkit that will speed up the process significantly.

First, when you've made a selection that goes beyond the edge of your screen, sometimes you want to quickly jump to the beginning and end of the selection to make sure you've got the right audio selected. As it happens, Avid has provided a few single key shortcuts that will do just that.

However, in order to use single-key shortcuts in Pro Tools, you have to be in what Avid calls Command Keyboard Focus. It's that little AZ box in the upper right corner of your Edit window.

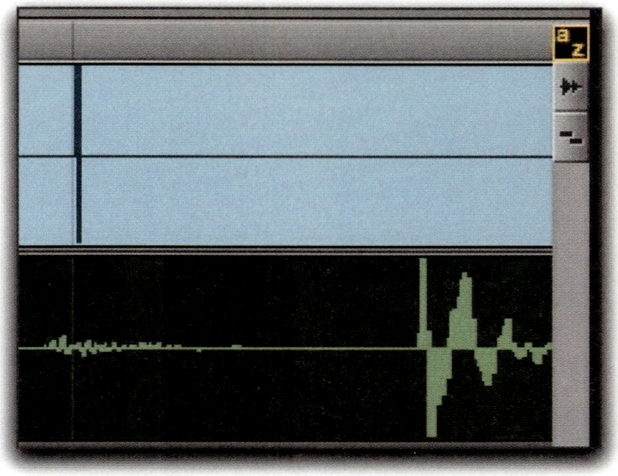

Figure 3.21
That's the Little Bugger, Right There.

When it is engaged, your keyboard will do many magical things that will turn you into a Pro Tools ninja. Turn it on if it's not engaged, then make a selection, and toggle between your Q and W keys. You'll no doubt notice that it transports you directly between the beginning and end of your selection, so you can expand or contract it as needed. Very cool.

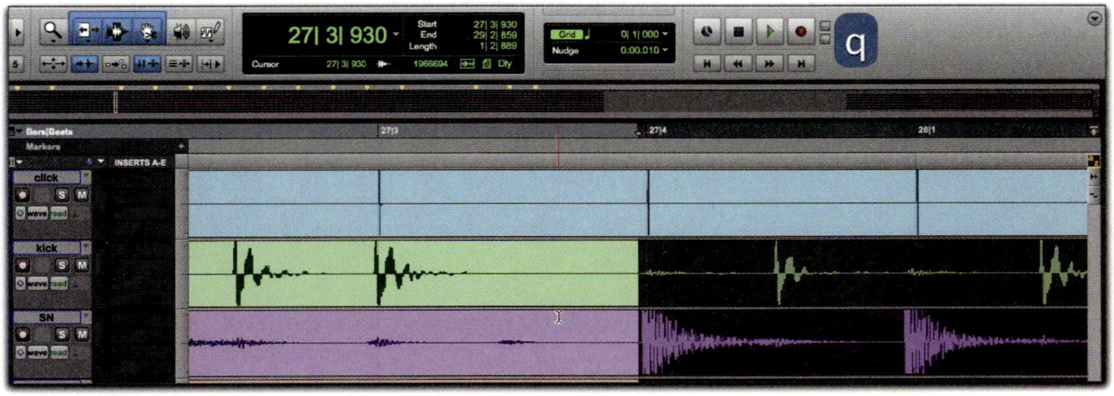

Figure 3.22

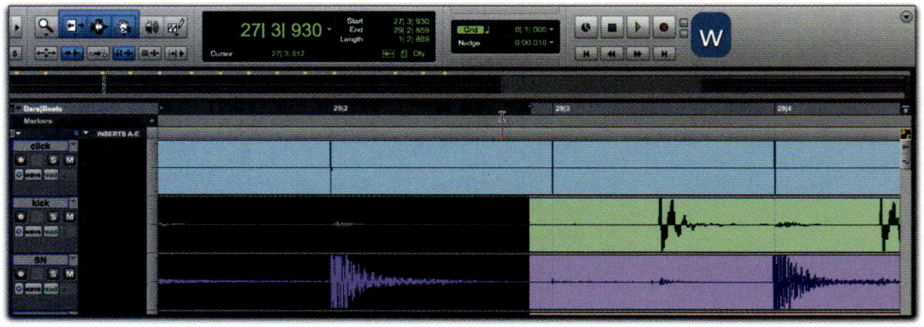

Figure 3.23 Use Q and W to QWickly (hmmm … That's Incredibly Clever … or Lame) Toggle between the Beginning and End of Your Selections.

We'll reference a number of these Keyboard Focus shortcuts over the course of the book or go straight to www.mixcoach.com to get a complete list if you're ready to jump ahead.

The Secret to Great Mixes

Many moons ago, I worked at a major pro audio retailer in the United States. Every day, I talked to dozens of aspiring engineers who believed that they were eternally a great piece of gear away from professional mixes. The fact is that while great gear is a huge component of great mixes, the lions share is the million tiny decisions that accumulate into a killer track. When you're done with this track, no one will ever wonder about how the drummer nailed such a rock solid performance, but it took your hours of edits to make it really shine. Stick with it.

How to Always Have a Job in the Music Industry

Over the last 10 years, I've taught at two of the world's largest audio production universities and one of the smallest. In that time, I've observed the common traits among those few students who went on to work on major, multiplatinum records receive Grammy nominations, start successful entrepreneurial recording careers, and equip studios for the largest of corporations. The secret of their success can almost always be attributed to one guiding principle: if you do your job in such a way that it makes your employer look better, you will always be on the first call list.

It can be applied in every situation. Students who were best prepared for class always stood out and received the first recommendations for the best internships, live gigs, and job openings. The studio assistants who made the engineer look better (by anticipating problems or preemptively volunteering to go get lunch) became the first calls for future sessions and on and on.

Following the logic of "The Secret of Great Mixes," it may not feel like these thousand little edits are making a significant difference in and of themselves. The fact is, though, it's the cumulative effect of the thousand decisions, which makes all the difference. When you finish, the goal is not to have anyone say "wow, great pocket! I don't even hear the edits." Or "nice tuning on that vocal!" Rather, while banging their head, the ultimate compliment for a mixer is "wow! great track! these guys rock ... "

So, keep that in mind as you spend the next few minutes prepping the drums. Roll down through the song, listen for trends and drifts over the remainder of the track, then go in and use your skills to nudge, or quantize them closer to the beat. After you're done, we'll use some new software tricks to take this track to 11.

Drum Transient Quantizing

Beyond the manual pocketing method we've just gone over, there's at least three other useful shortcuts for rapidly tightening up percussive performances in Pro Tools (say that three times fast): (1) Chop and Quantize, (2) Beat Detective, and (3) Elastic Audio.

The first method we're going to look at is most useful on a section of audio that has clear transients and a steady, predictable beat. Since I'll assume you've already gone through the track and nudged all of the sections close to the beat, this technique should be most effective here at the end of the song. Go ahead and select from about 127|4 to the end of the song.

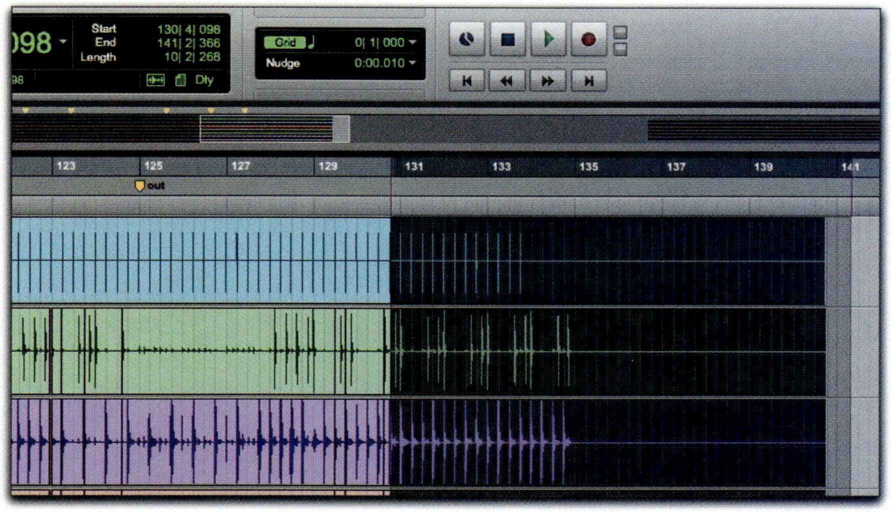

Figure 3.24
A Simple, Straightforward Beat That Just Needs a Bit o' Tightening.

Once you've got it selected, separate it (Mac: Cmd+E, Win: Ctrl+E), then go to Edit>Separate Region>At Transients.

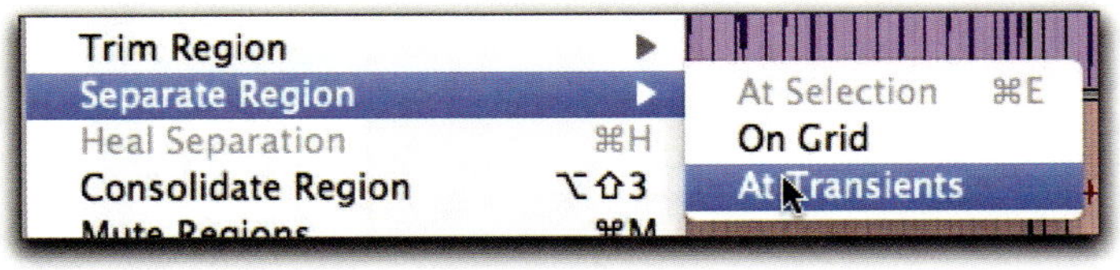

Figure 3.25 "At Transients" Is the Lesser Known Brother of the Popular "Separate At Selection" Command.

A dialog box should pop up asking you how many milliseconds you want to create the cuts *before* the transient. For example, if you were using this to cut out a bunch of bass notes, which have a relatively longer transient than a drum, you might choose to separate 15–20 ms *before* the transient, so you're not cutting off the ramp up of the note.

However, with drums, we can often go with a Pre-Separate Amount of 0–5 ms depending on whether it makes this particular drumkit sound like the front of the notes have been clipped. We'll start with 0.

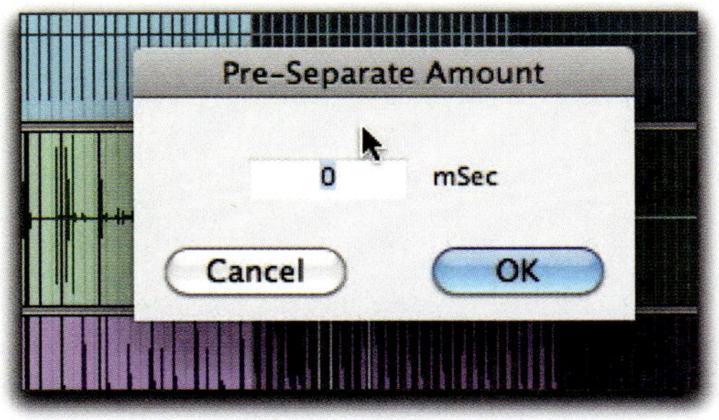

Figure 3.26 Cut It Right On the Transients.

Voila, Pro Tools does the heavy lifting of going through your drum tracks and chopping every one at the transients.

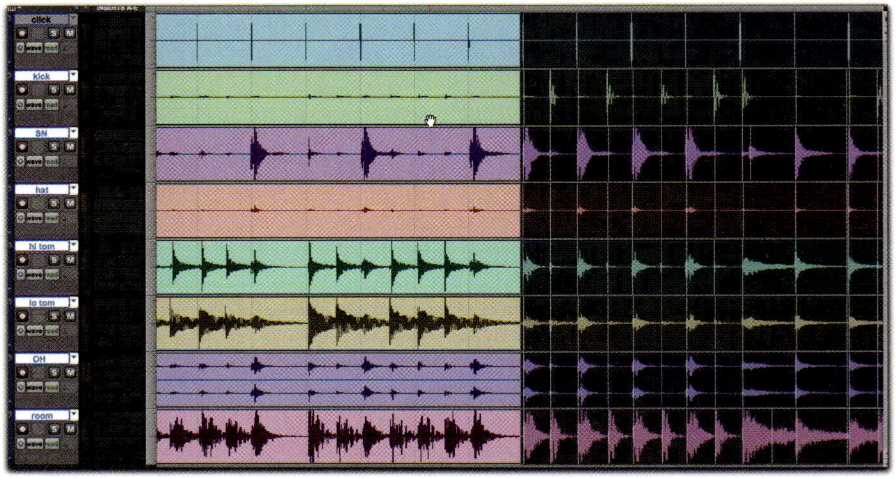

Figure 3.27 Well *That* Was a Lot Faster than Tabbing to Everyone.

Now that they're all chopped up, do you remember the Grid values up above? If it's still set to quarter notes like it was earlier, hit Command+0 (Win: Ctrl+0) to get a quick visual of what will happen to all those regions.

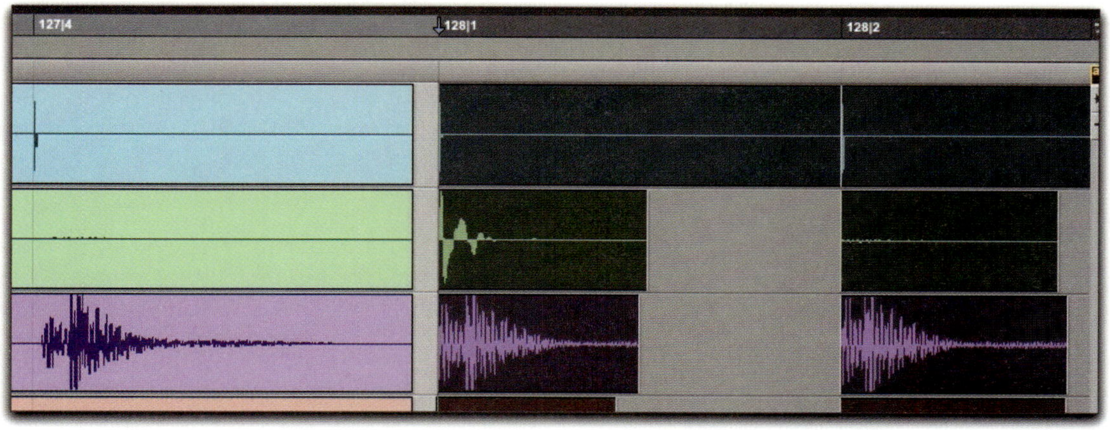

Figure 3.28 Oops. That's Not What We Wanted.

You should see the regions all get pulled to the nearest quarter note, which of course is not what we want. For that matter, even quantizing them to eighth notes, while it puts things in the proper time, is still not an effective way of editing multitrack drums because it doesn't take into account anything but the transients, even across multiple tracks. As a result, the regions can get shifted out of time with one another, creating a phasey sync disaster.

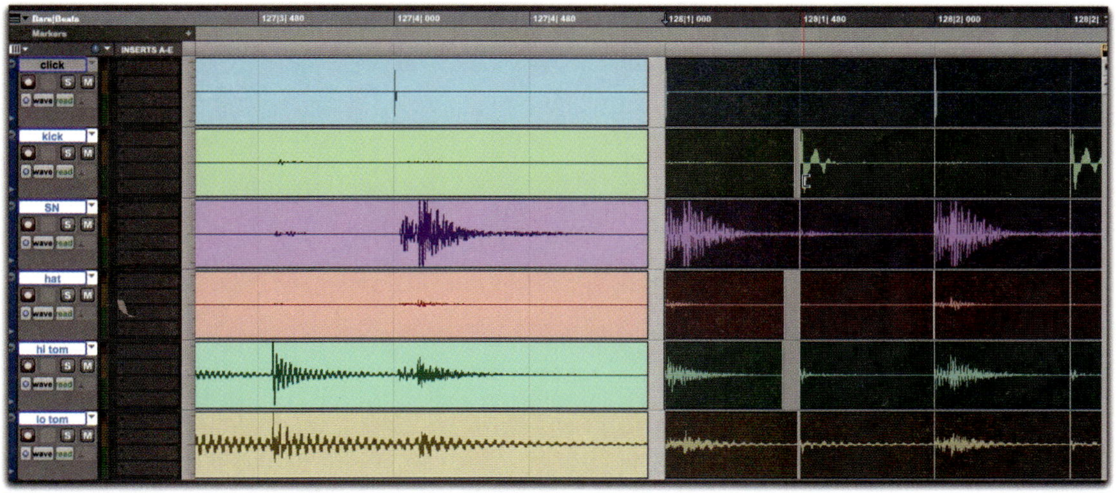

Figure 3.29 Yeah, Still Not What We Wanted.

This technique is especially effective on any *single* track that needs to be quickly lined up, like a shaker, sound replaced snare, tamborine hit, etc. but *not* for multimic instruments.

So let's hit our Edit>Undo command (Mac: Cmd+Z, Win: Ctrl+Z) twice to both put the audio back where it was, as well as undo the transient editing.

Beat Detecting Multitrack Drums

So, if the quantization method isn't effective for multimic drums, then what is? Let's go look at our old friend Beat Detective.

Start by making a selection from around bar 4 to bar 14. Then, hit Cmd+8 (or the Event menu) to bring up Beat Detective. For those of you who are new to shortcuts, Pro Tools requires you to use the numeric keypad on your keyboard for shortcuts with a number.

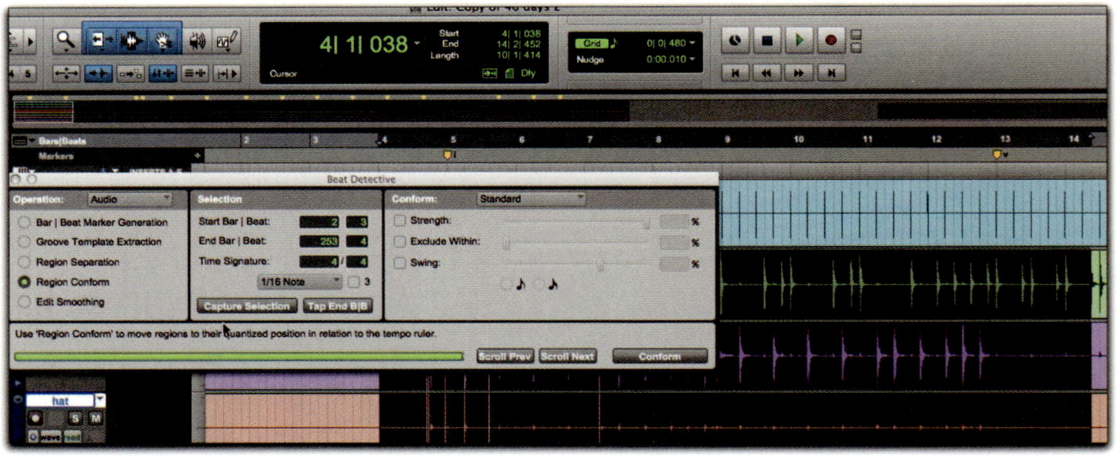

Figure 3.30 Don't Be Afraid. It May Look Confusing, but We Can Bend Beat Detective to Our Will.

First things first, click the Capture selection button to load the selected audio in to Beat Detective. Now, toggle to the region separation option under the operation area. You should now see purple lines across all of your drum tracks indicating detected beats across all of the multimic drums.

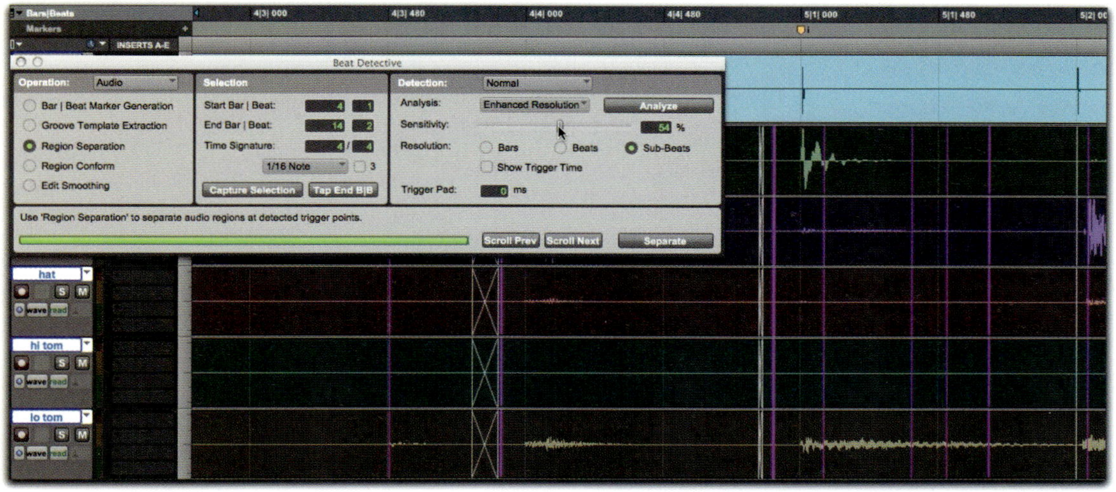

Figure 3.31

Initially, there may be far too many detected transients. So, before we move on, click the Analyze button and drag your sensitivity slider all the way to the right and all the way to the left. You will notice as the sensitivity approaches 100%, there are far too many transients detected. At 0%, there are no transients selected. At around 29%, you should notice the sensitivity begins to line up very well with the selected drum transients.

So at this point, we could use a separate button to chop up our drums. But, if we try that, Pro Tools will tell us it cannot separate an existing crossfade.

This happens on occasion when you are editing. For example, if you have been quickly editing, or punching in a variety of parts and making crossfades to hide the edits during tracking, when you get ready to use the detective, you need a quick way to delete all of those crossfades from the tracks. That option is located under the Edit menu>Fades>Delete. Voila, no fades.

(a) (b)

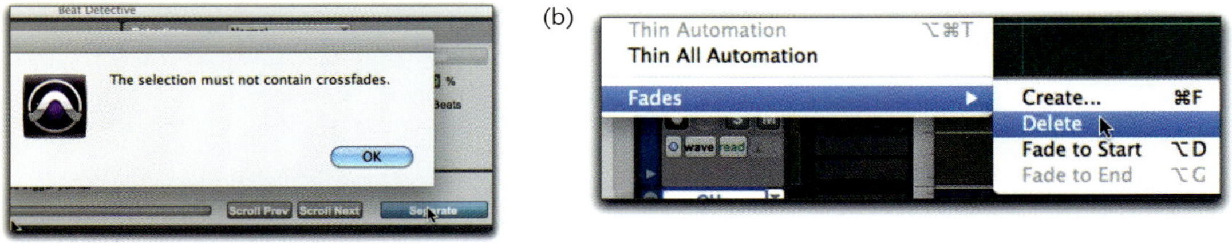

Figure 3.32 (a, b)

Now, hit the Separate button and you will notice that, similar to the "Separate Region>At Transients" option we used earlier, the audio is broken up into independent regions. This time the difference is that the regions all start and end at the same time, keeping each sound phase locked with the other tracks in the group.

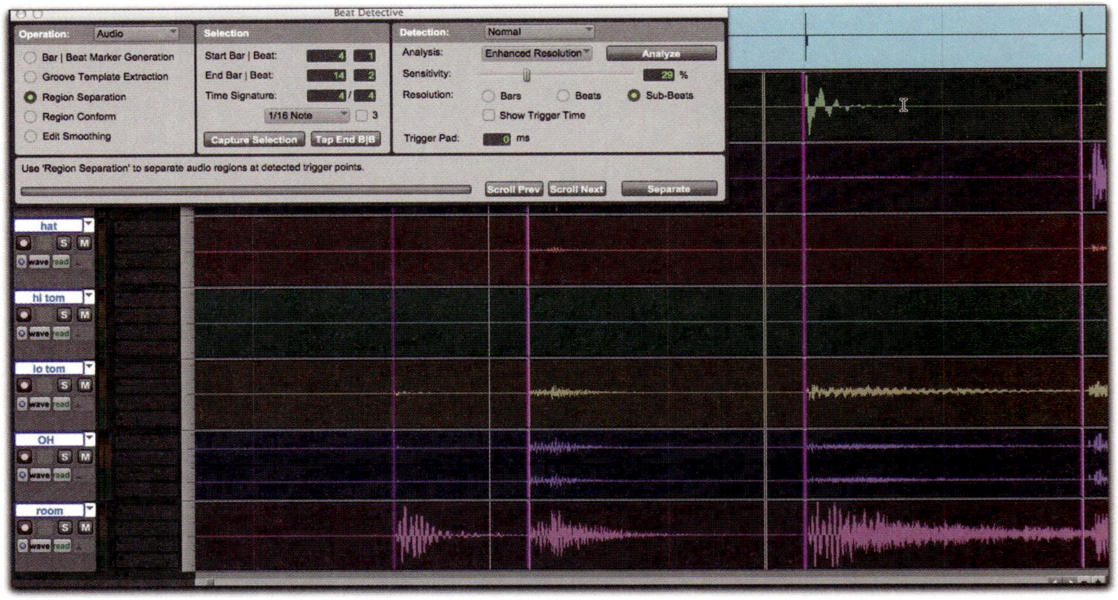

Figure 3.33

Conforming Drum Edits to a Grid

That's much better. Now, we can either use the Region Conform Operation in Beat Detective to quantize all of the selected regions to the given time signature and tempo or we can go back to our old standby Quantize to Grid value (Alt+0).

The benefits of doing it in Beat Detective are that it's already open and ready to go. It also has a few options for choosing the Strength, Exclude Within, and Swing, which will allow you to quantize less (with a Strength of <100%), and Exclude notes that are already within a few percent of the correct time, or even add a swing feel to the regions.

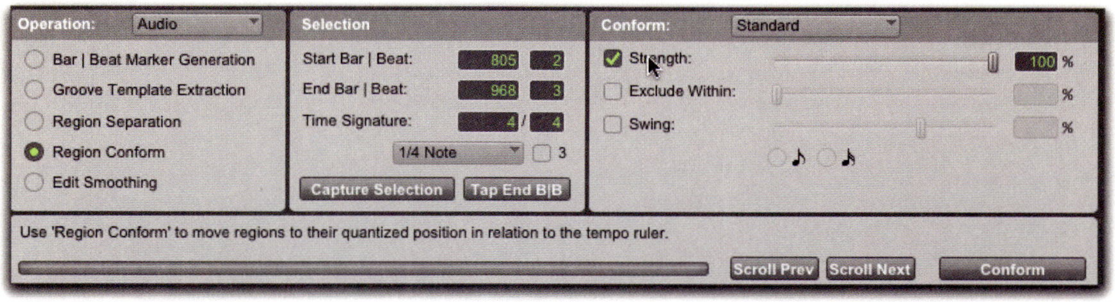

Figure 3.34 We Could, but We're Not Gonna.

In our case, we're trying to make these drums rock harder and more in time, so one Quantize keyboard shortcut will automatically snap all of these regions to the nearest sixteenth note. So, select all the newly separated regions, set your grid to 1/16th notes, and hit Alt+0.

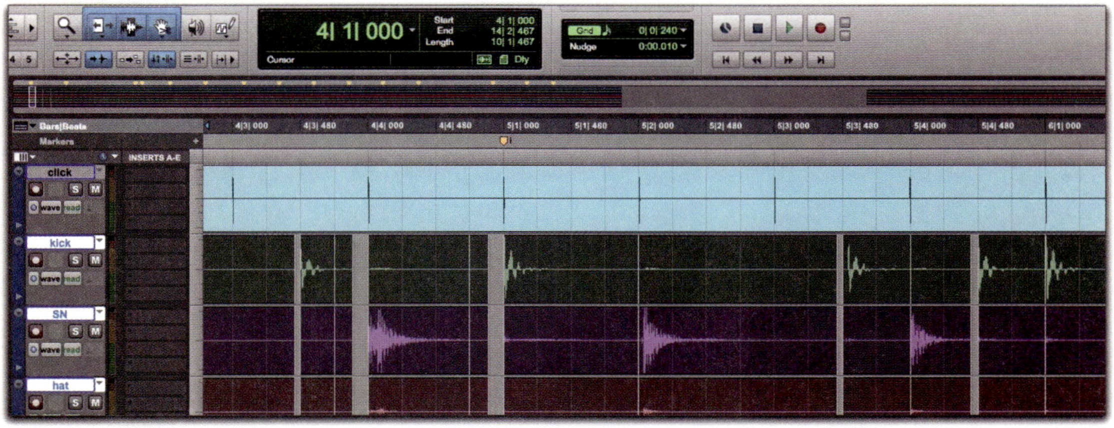

Figure 3.35 Well, They're in the Right Place … but There Are Some Ugly Gaps in the Audio.

Smoothing Audio Gaps

"But what about all those gaps left in the audio?" I hear you scream from your comfy chair. You are observant, grasshopper. It's Beat Detective back to the rescue with its groovy Edit Smoothing Operation. Open it up, switch to Edit Smoothing, then set the Fill and Crossfade option to around 5 ms. Click smooth, and you will watch the green bar effortlessly trim all regions back to fill in any created gaps, apply a 5 ms crossfade to hide the edits, and even leave you a handy little green sync marker next to each transient, to show you that the sync marker is the quantizing reference, rather than the left edge of the edited region.

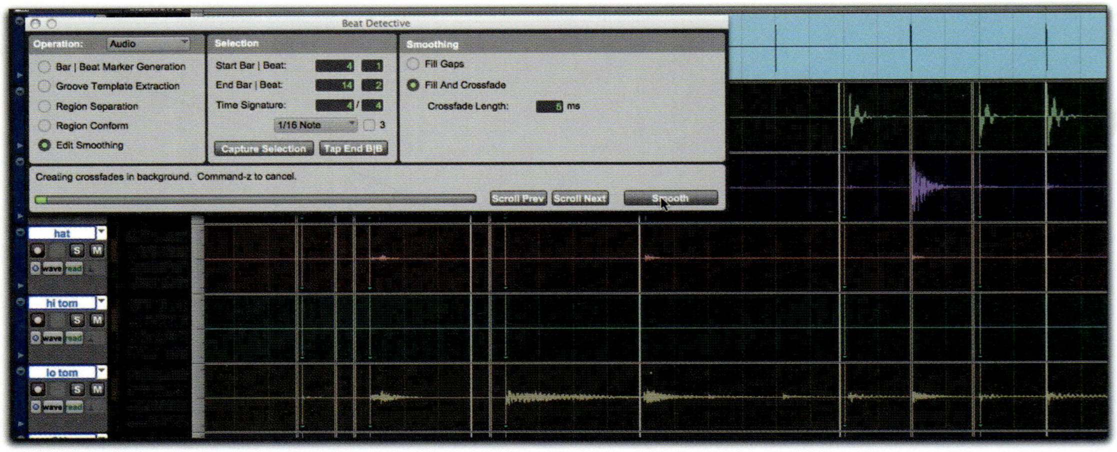

Figure 3.36 It's So Pretty.

Go ahead and give your now-perfectly-timed drum selection a listen, you've earned it.

Method 3: Elastic Audio

Now that you've got a pretty good feel for how to rough-in a drum track, as well as how to use the Quantize function and Beat Detective to chop and move the beats around, let's use a relatively new function that DAWs like Pro Tools, Logic, and even GarageBand have implemented in the last few years, Elastic Audio.

The quick and dirty explanation of Elastic Audio is that it's a relatively new function that operates behind the scenes in Pro Tools to time stretch or compress individual notes in audio tracks for the purposes of timing correction or effect.

In other words, if you switch on Elastic Audio on a track, Pro Tools will analyze the audio content (similar to the way it searched for transients with Beat Detective) looking for rhythmic and tempo information. Then, when you are viewing the new Elastic Audio (hereafter, shortened to EA) information, Pro Tools will allow you to perform previously impossible time changes on audio (Quantize, Swing, Speed up or Slow down, etc.), just like we've always been able to do with MIDI. It's an incredibly powerful function that works well when used properly. However, used improperly, it can create audible weirdness, glitchy sounds, and metallic effects on acoustically recorded tracks.

So, our objective when using it on drums is to figure out how much we can use EA to speed up the process of time correction, without making our drums sound like glitchy mush. Let's dive in.

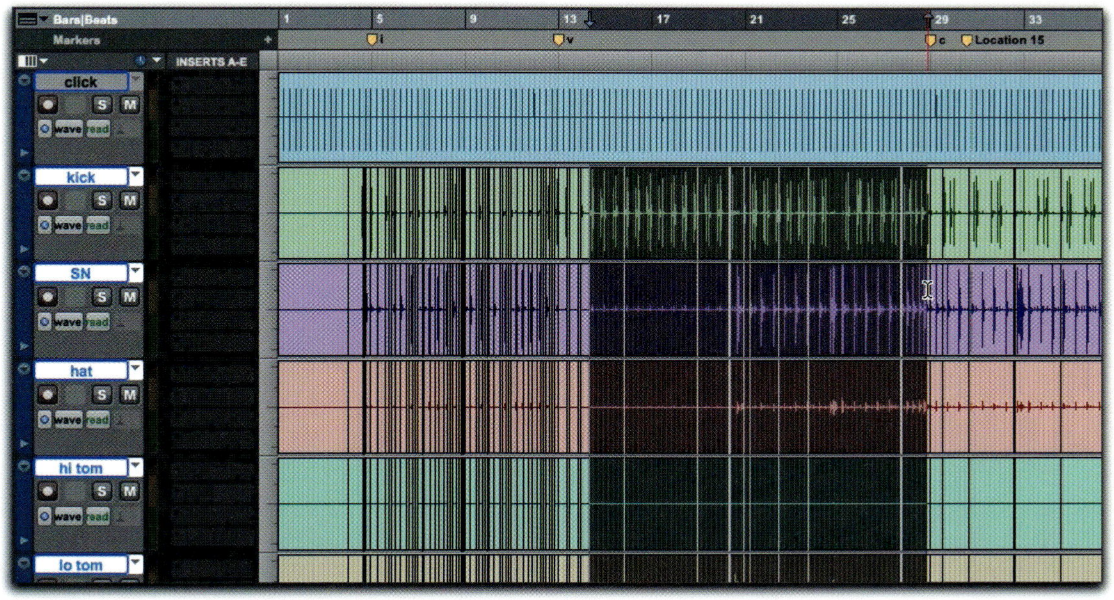

Figure 3.37

Since we've already roughed in the drums and Beat Detected the intro, let's start by selecting the first verse of drums to Elastic Audio (can it be used as a verb? Hmmmm. I like it).

Make sure your drum group is still turned on because we definitely don't want to shift any of the drums separately in time from the others. Now, click on the little metronome icon in the track headers below the Mute button and choose the Rhythmic EA algorithm.

(a)

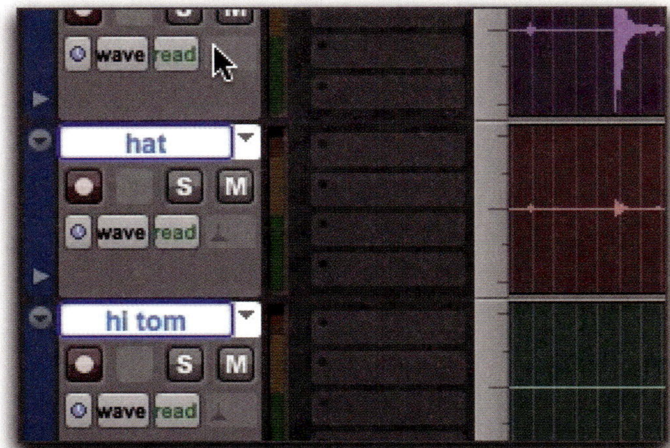

(b)

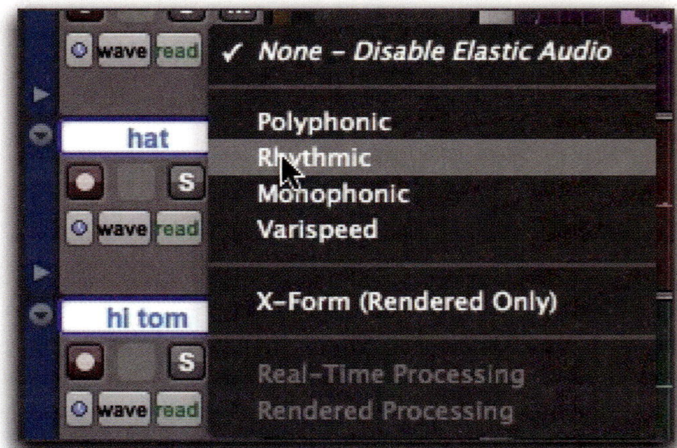

Figure 3.38 (a, b)

Different algorithms are useful for different types of sounds and instruments. For example, Polyphonic is designed for instruments with more than one tone playing simultaneously (e.g., Piano, guitar, etc.), where Monophonic is for single note instruments (Bass, Vocals, etc.). The Rhythmic algorithm attempts to identify and avoid stretching the transients of the note, so they don't lose that percussive attack.

After you've chosen Rhythmic, your tracks will go gray for a minute as Pro Tools attempts to analyze them for transient and rhythmic information. When they are finished analyzing, switch your view to Analysis to see the little white transient lines that Pro Tools thinks it found.

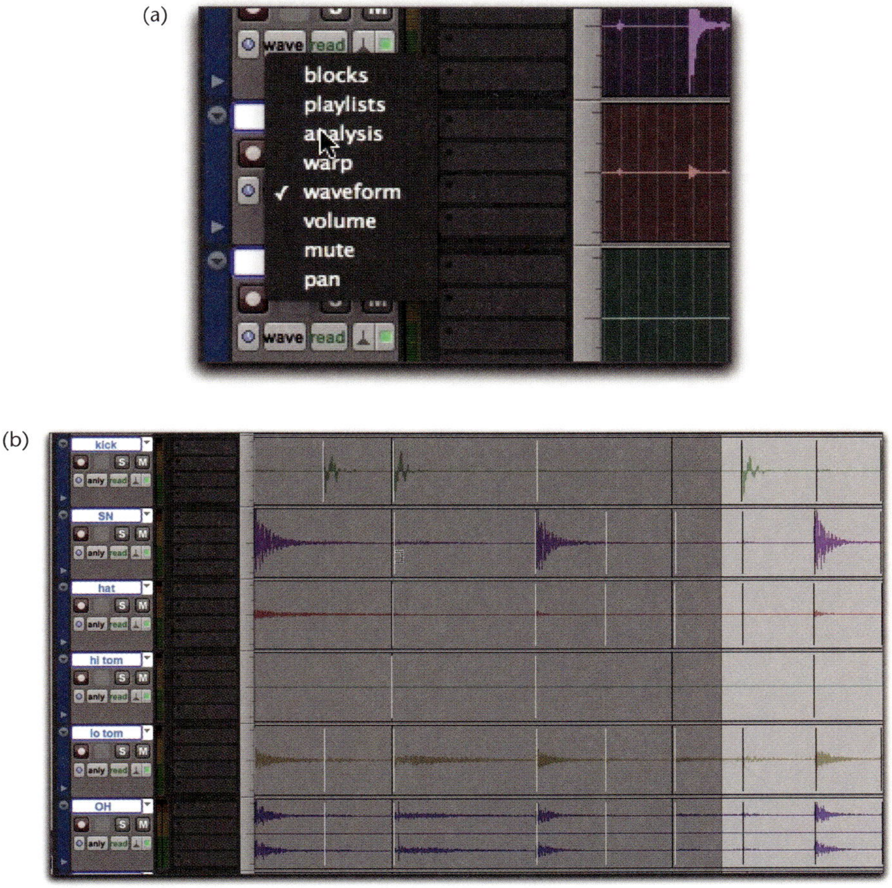

Figure 3.39
(a, b) Choose Analysis, Warp, Waveform, or Other Views for Accessing Editing Functions on Your Tracks.

If needed, the Analysis view can be used to correct any false transients PT may have picked up, however, the bulk of our EA work will be done in Warp view. Switch over to Warp view now and let's explore the tools.

Just as we did with playlists, I want to first show you how to avoid getting yourself in trouble with Elastic Audio.

Getting into Trouble with Elastic Audio

The challenge with all of these great tools is that they can be so powerful, and far-reaching in your session, that it is very easy to make what you think is one small timing change, which has a ripple effect of moving things out of time with the rest of the session. Anyone who's ever accidentally made a bunch of edits in Shuffle mode knows what a hassle it is to re-align their audio once they've exceeded their maximum number of Undos.

In the same way, if we just start grabbing and dragging Warp markers around, Pro Tools will also stretch or compress the entire rest of the track, throwing everything out of sync.

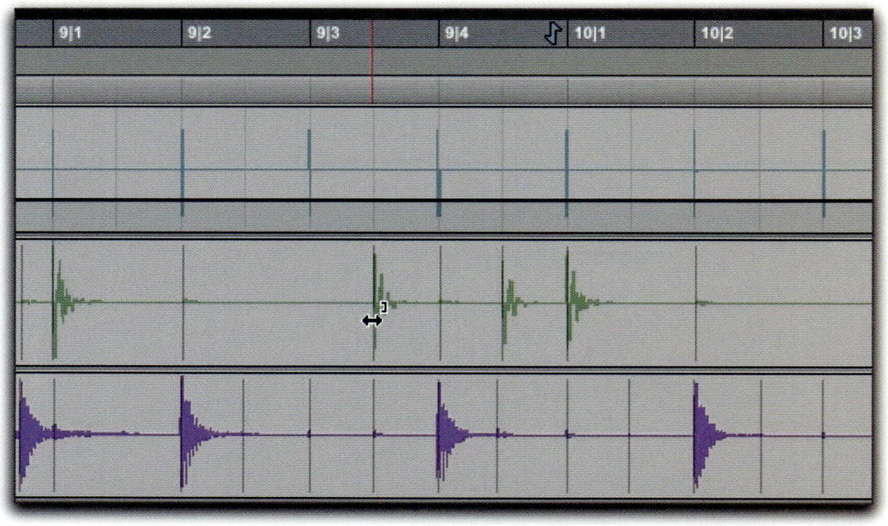

Figure 3.40

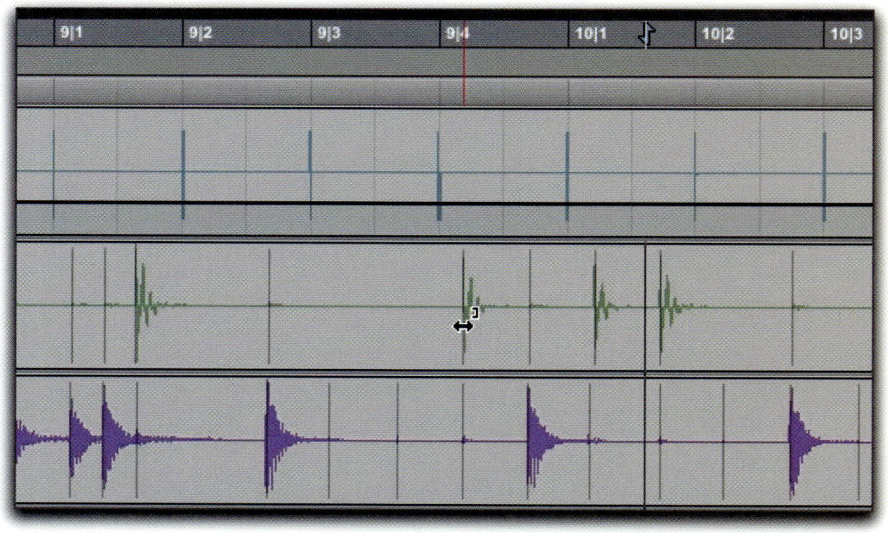

Figure 3.41
Without Boundaries, the Entire Drumkit Moves When You Drag Any Warp Markers.

That is, unless we set up boundary markers first.

So to start, select the drums in your verse, from approximately 13|3 to 29|3. Then, hovering over an Analysis line around 29|2, hold down the Shift key, and click to create a trio of Warp markers.

(a)

(b)

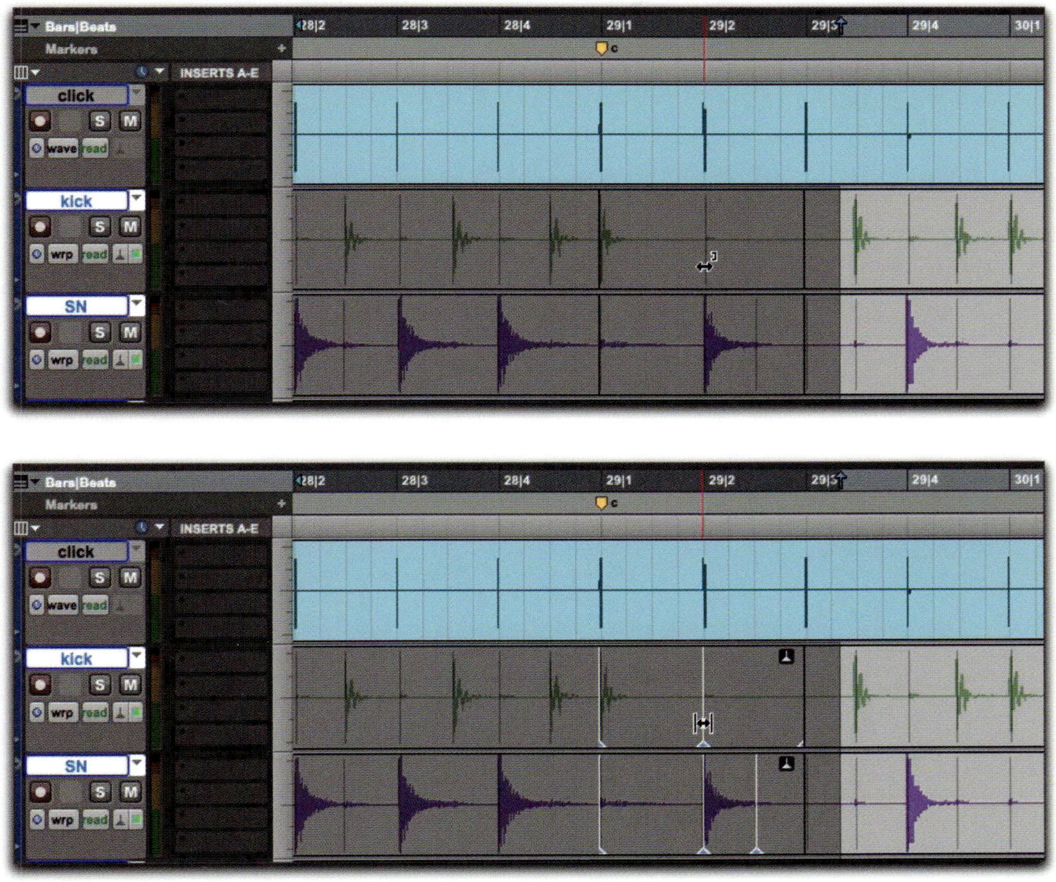

Figure 3.42 (a, b) Setting Warp Marker Boundaries.

Just like that we've created a set of Warp markers that will act as a boundary whenever we drag any Warp markers to either side of it. Now, use your Q key to quickly jump back to the beginning of your selection around bar 14 and Shift+click on an Analysis point to set a Warp marker boundary.

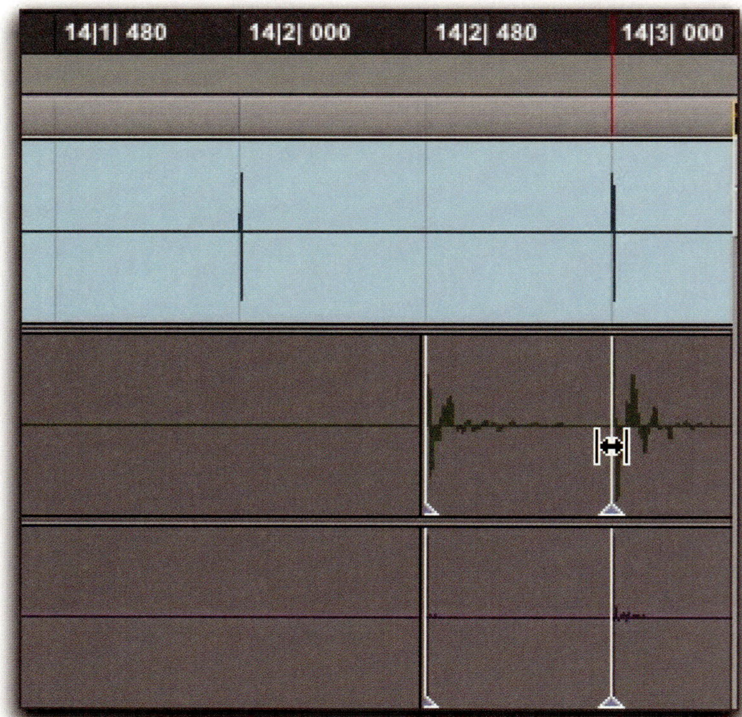

Figure 3.43
More Warp Marker Boundaries.

Now we're set. Similar to when we used the first two drum editing methods, but without the hassle of chopping up the audio and then fixing all the gaps, we can now go through and manually drag the markers on the Kick and snare tracks to their nearest beat or autoquantize them.

Let's do both, for two reasons. One, when you use any "auto" function in software, even with all of the modern advances, the software still can't *hear* the performance or know what the drummer was *trying* to do. As a result, if you tell it to quantize to the nearest sixteenth note, it will blindly follow orders, even if the drummer throws in a special fill, or changes time signatures. Then, the resulting Elastic Audio will no longer fit the other instrumental performances, which is obviously bad.

Assuming you did a reasonable job during the rough-in phase, all of the drums in the verse section should be close enough to their proper beat to line up well with a Quantize command.

Quantizing Drums with Elastic Audio

So, select all the audio between your boundary markers, make sure you're in Warp view, and hit Alt+3 (Or Event Menu>Event Operations) to bring up your Event Operations window. Set your options to Note On, Preserve Note duration, Elastic Audio Events, and 1/16th note grid. Finally, click Apply.

(a) (b)

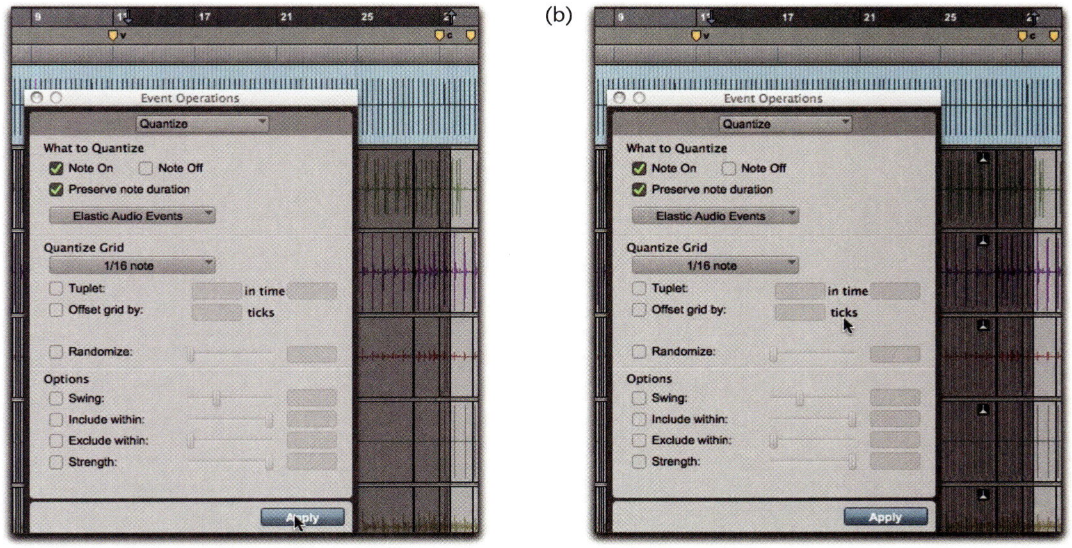

Figure 3.44 (a, b) It's Almost Too Easy to Fix Modest Performances Nowadays.

Can you believe how easy that was? Sure, the drummer was talented and had to be reasonably close to the beat to begin with, but with a simple analysis and a click, we're able to dramatically enhance the timing to sound like a talented drummer, *on his best day*. It's a magical thing.

Now as you listen down to the new drum tracks, inevitably you'll find a few notes he might have thrown in that got moved to the wrong notes. Fixing them can be handled in a few different ways depending on the situation.

One, if the note is in the middle of a long string of unedited audio, it's as simple as switching to Grid mode, holding the Shift key, and click-dragging the Warp marker on the errant note to its correct beat. Go ahead and find a note that isn't properly quantized (I found one around bar 21|1), hit the F4 (function) shortcut key on your keyboard to switch to Grid, hover over the Warp marker, and click and drag it to the correct beat.

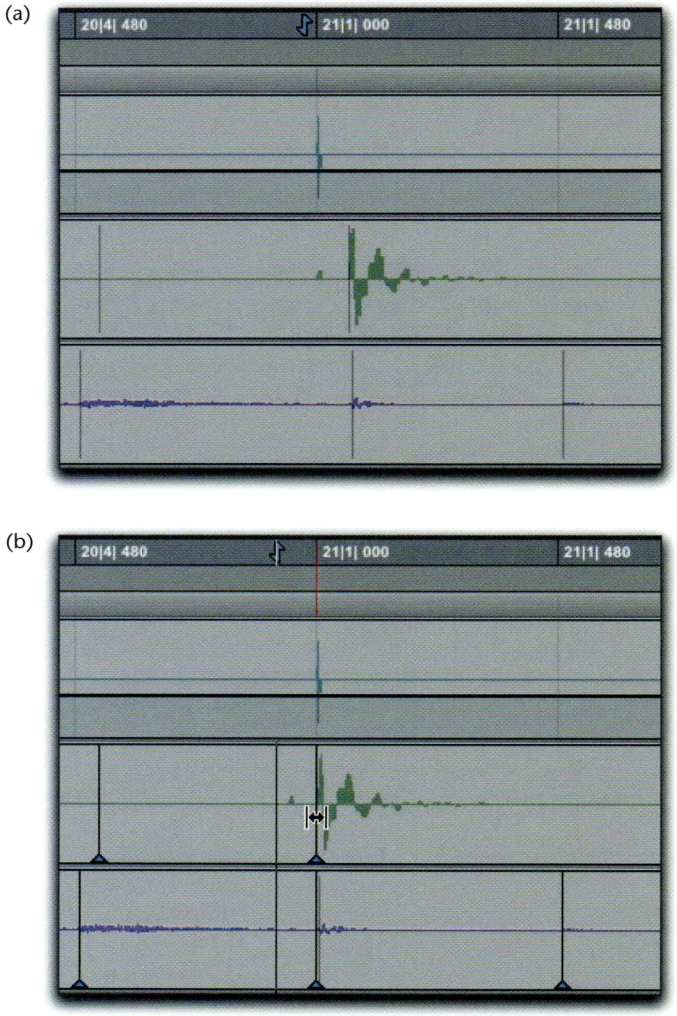

Figure 3.45
(a, b) The Late Green Kick Note Can Be Manually Quantized with a Simple Shift+drag to Line It Up with the Beat.

Combining Drum Editing Methods

This will work for 85% of your edits. One situation where it may have problems is if the note is misplaced due to being too close to an audio edit in the waveform. If so, you can also just switch back to your Waveform view, select the entire note, separate and nudge or quantize it to the proper beat.

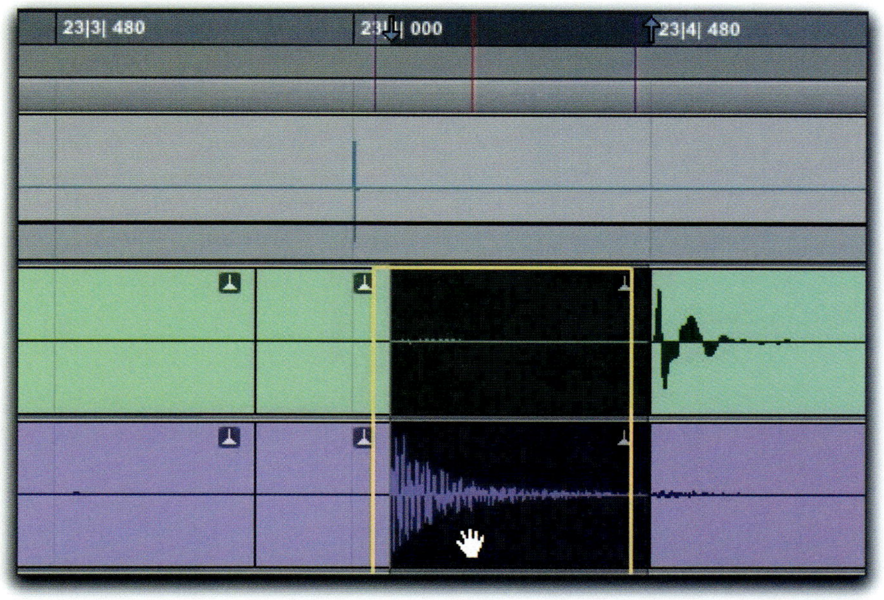

Figure 3.46 Just Flip into Waveform View and You Can Combine the Best of Elastic Audio with the Best of Manual Pocketing.

> Want a shortcut to avoid a bunch of mousing around? Go back to our awesome fat finger trick (Control, Alt and Command/Windows) and tap the right- and left-arrow keys on your keyboard. This will toggle all of your tracks between the various views and let you jump between Warp and Waveform views with a single keystroke.

Another problem you'll run into when using Elastic Audio on drums is that cymbals that get stretched too much can take on an odd, phasey sound. After finishing our drum pocketing, I hear one of these at bar 129.

When working with rock music, one advantage is that it's often simpler to copy a usable note from another section and paste it over the offending note, than it is to rerecord or re-edit the glitchy sound.

So, rewinding a bit into the track, I've found another similar kick/cymbal downbeat at bar 115. Let's select just the first beat and copy it with Cmd+C (Win: Ctrl+C).

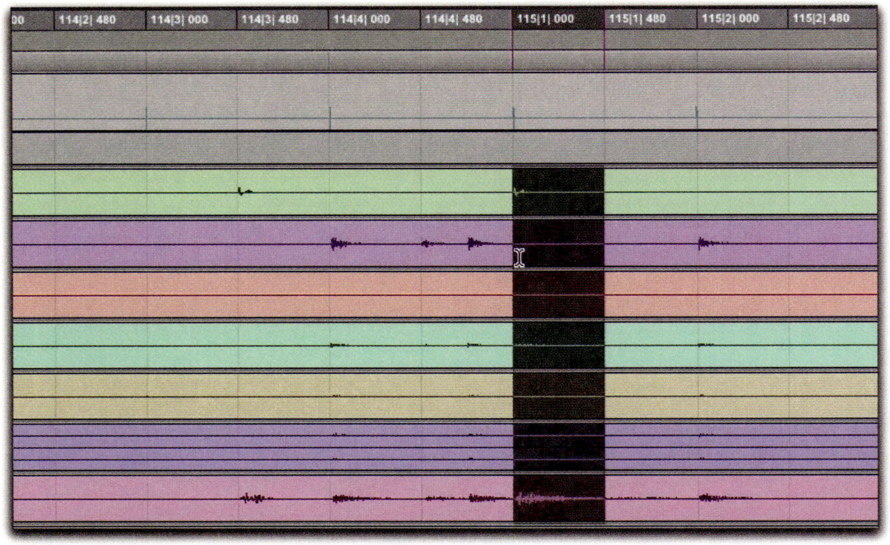

Figure 3.47 Copy a Good Cymbal Crash from an Earlier Part of the Song.

Now, roll forward to bar 129, place your cursor (using Grid mode) right on the beat, and paste (Mac: Cmd+V, Win: Ctrl+V) the good audio over the bad. Use the Trim and Crossfade tools in Slip Mode to move the end of the edit over near the snare hit at 129|2 to hide the edit, and you're ready to go.

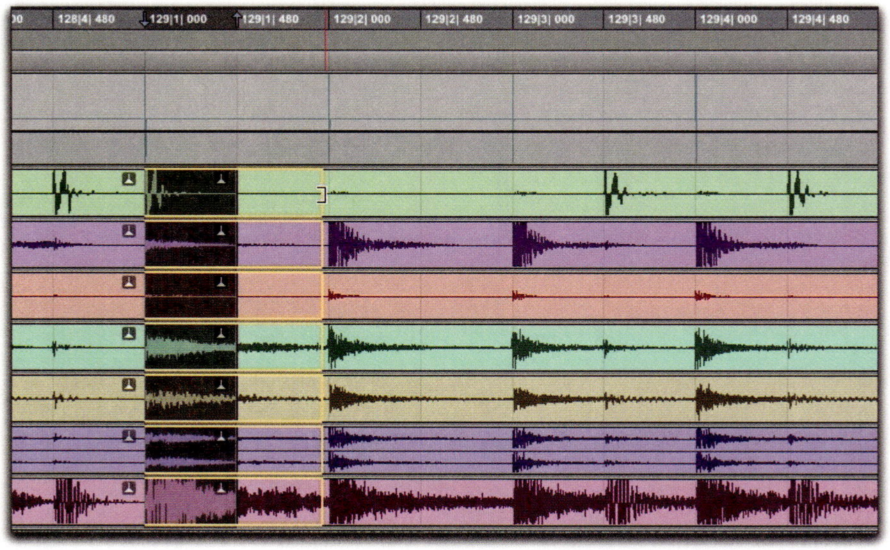

Figure 3.48 Trim Your Drum Regions Near the Next Beat and Add a Crossfade between Them to Hide Your Edit.

Finally, especially while you're practicing these techniques, if you find yourself hopelessly messing up a section of the track, and needing to get back to the original drum performances, simply switch over to your unedited, original playlist, select and copy (Mac: Cmd+C, Win: Ctrl+C) the section you've screwed up, switch back to your editing playlist, and paste (Mac: Cmd+V, Win: Ctrl+V) it in. Ahhhh. Now you can try again, secure in the knowledge you've got the originals just one click away.

Now it's true that editing the drums is often both the longest and most tedious part of the mixing process. Depending on the tightness of your drummer, it can be a reasonably quick or painstakingly time-consuming job.

That said, if the project and budget call for it, you need to have every one of these techniques in your Mixing Toolkit. As we've already explored, the key to staying employed in the music industry is to always make the people paying you look and sound their best. Having every trick up your sleeve is key.

In This Chapter

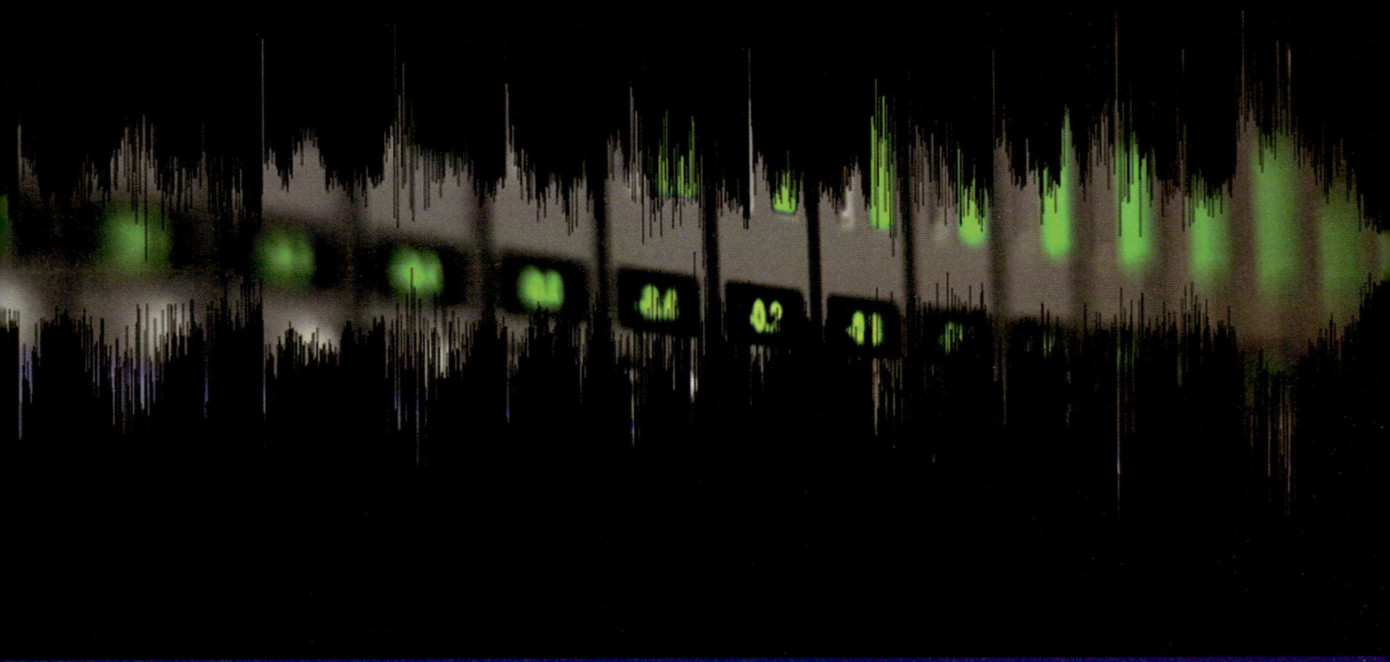

4

Bass Guitar

Setting Up Your Bass Pocket

Now that you've spent a little time practicing multiple methods of pocketing drums and percussion, let's dig in to the bass guitar. Hold your Shift+Alt shortcut down and click+drag up between any of the drum tracks to shrink them to minimum height, then turn on your bass track and resize it to make it large enough to see. Groovy. Give it a listen.

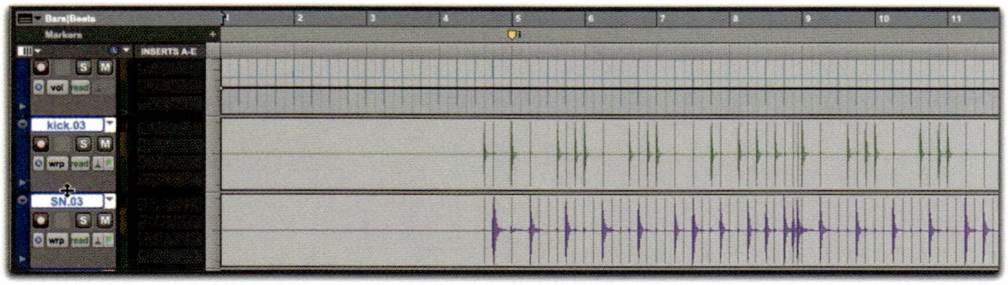

Figure 4.1 Resize Your Drum Tracks.

Figure 4.2 Listening Down to the Bass Guitar.

© 2011 Nathan Adam. Published by Elsevier, Inc. All rights reserved.

Not bad. Overall, his performance is in the groove, but as I'm sure you're hearing, once you've started pocketing the drums, all of the other tracks may start to feel a little *more* out of time. It's really a process that requires commitment for the entire project, so let's go ahead and pocket the obvious notes, and then you can translate those skills into pocketing the rest of the track.

Elastic Audio on Bass

During the intro, it feels like he gets out of the pocket a little bit. Of course, if the band has the budget, now would be the time to work through the entire track using either manual pocketing or Elastic Audio. Using Beat Detective is generally more effort than it's worth on a musical track, as the transients aren't as easy to detect inside sustained strums and chords. So to start, let's go ahead and turn on Elastic Audio and start by fixing a few obvious notes.

Because the bass guitar is generally playing one note at a time (bass players can't handle any more than that … ba-dum, crash!), we will use the monophonic algorithm (mono = 1, phon = pertaining to sound). Polyphonic is more applicable to multitimbral instruments such as piano or acoustic guitar.

Pro Tools is going to analyze the bass guitar track, and when it is finished, we can start pocketing the notes.

(a)

(b)

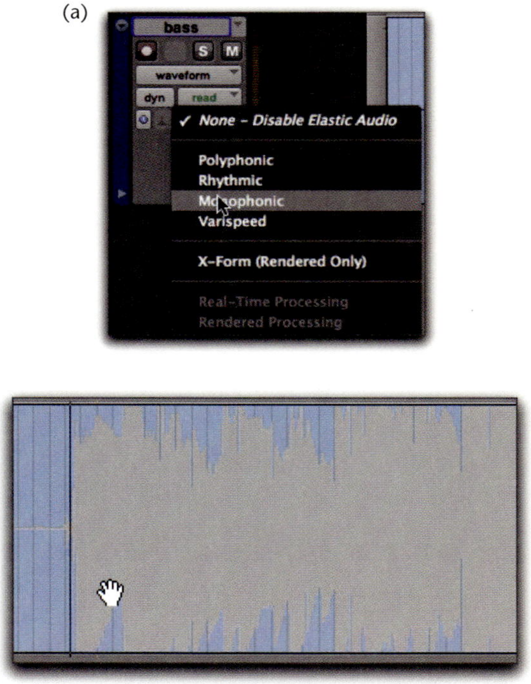

Figure 4.3
(a, b) Analyzing the Bass Guitar for Transients.

While we're waiting, it's worth noting that while the processes of recording, editing, and mixing used to be (and still can be) handled by multiple engineers, in the modern world it is often a lone entrepreneurial engineer handling tracking, editing, and mixing duties (and raise your hand if you've ever done the project mastering too … that's what I thought).

The fact is, whether you're cutting the tracks or you're hired to mix a preexisting session, a large part of getting a prolevel sound is ensuring the performances are as tight as possible. Otherwise, you will spend an inordinate amount of time making the low-end sound tight if the kick and bass aren't solidly together. Further, if the band isn't sufficiently tight (they often aren't) or they haven't hired another editor (they rarely do), then it falls into the modern mixer's lap to perform these functions before the mix to make sure it sounds great. As a result, I usually try to get my editing out of the way before starting the traditional "mix."

I want to be critical in other, musical ways when I start to mix, not thinking about timing.

Rabbit trail aside, switch to Warp view on the bass track, and you will see that a lot of the Warp markers are already in place ready for editing.

(a)

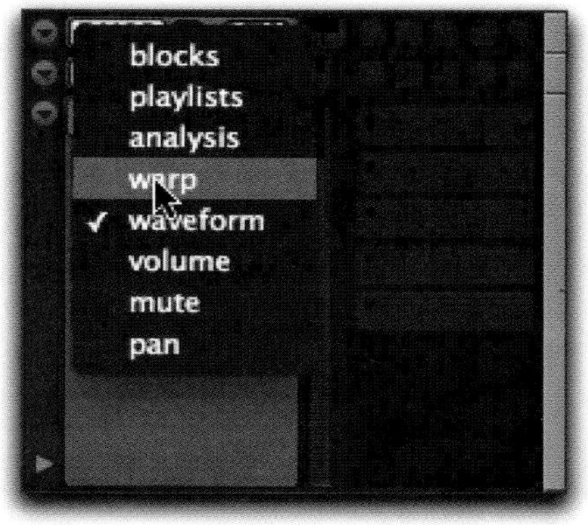

(b)

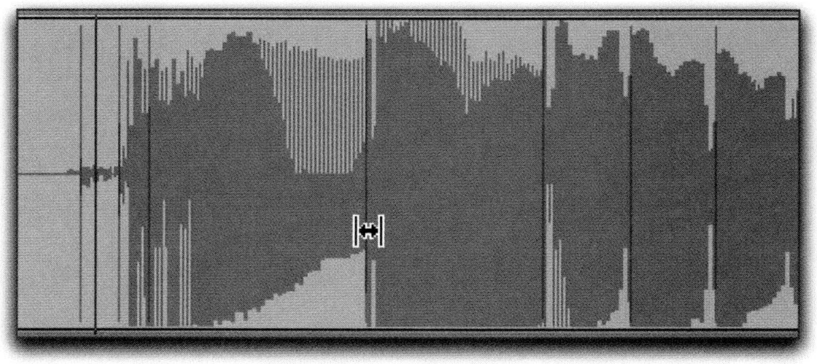

Figure 4.4
(a, b) Analysis Markers on the Bass Track.

Start to listen through now, identify notes that need a little nudge. Just like when we were editing the drums, by holding the Shift key, we can select the waveform and move an individual note without moving the entire track. Start by adjusting the first note to line up the transient of the first bass note with the kick drum at 4|4. Use the images below, and your ears, as a guide.

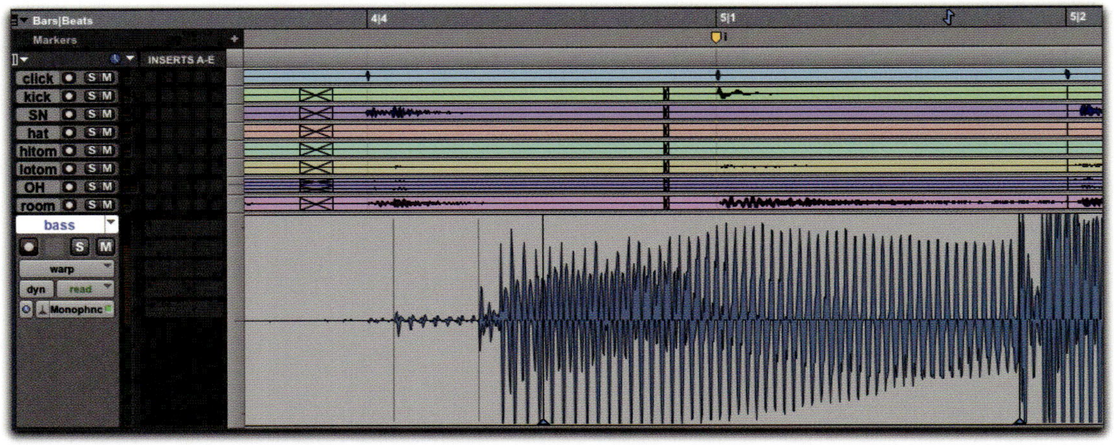

Figure 4.5 Bass Before Time Stretching.

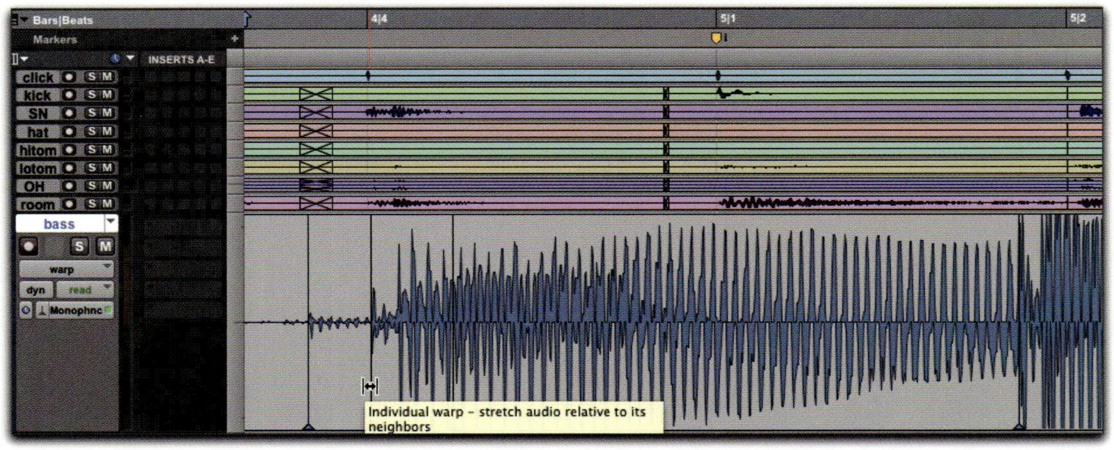

Figure 4.6 After. Now Your Bass Slides in Dead on with the Drum Flam.

Now you're probably noticing a visual difference between EA editing a bass versus the drums. Because of the sustaining waveforms in the bass, it can sometimes be really hard to see the Analysis markers to know where to drag.

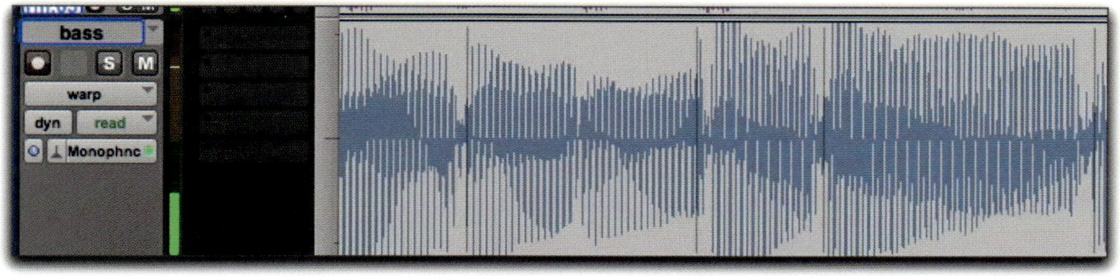

Figure 4.7 Where Are They Again?

No problem. Just adjust your waveform size so that you can see a good blend of Analysis markers and waveforms. You can either use the Waveform resize button or the shortcut Alt+Cmd+[ ] (left and right brackets; Win: Alt+Ctrl+[ ]).

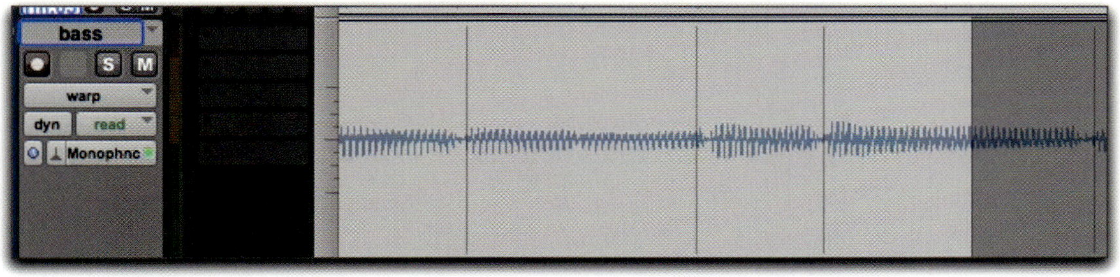

Figure 4.8 Ahh. There They Are.

Now, for this track, we are not going for a machine-like quantization sound on the bass guitar. We're shooting to keep a bit of live vibe in the track. On the flip side, if we wanted to, we can make a selection over a few bars (say 6|1 to 7|1) and just quantize the transients.

Quantizing Bass

Lest we shift the entire track though, as seen before in the drum section, hold the Shift key down, move to the end of the intro (around bar 7), and Shift-click a Warp marker, effectively setting a boundary for our upcoming quantize function. Now select between the section between bars 6 and 7, Open your Event Operations window, and quantize to eighth notes.

(a)

(b)

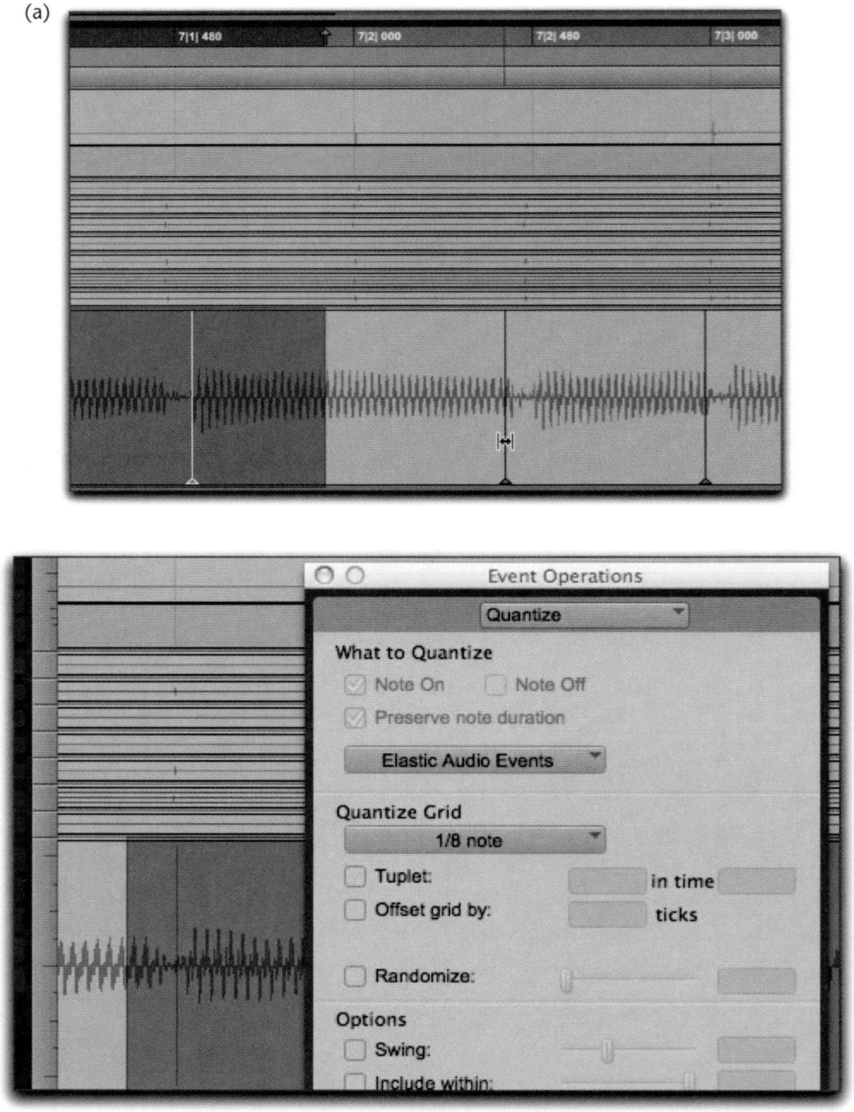

Figure 4.9
(a, b) Quantizing the Bass.

Kablam. All of your Bass notes magically snap into place, and we only need to listen through, moving individual notes more into the pocket.

The next one I hear arriving early is around 9|1. Shift+click the Warp marker and drag the transient until it starts at 9|2.

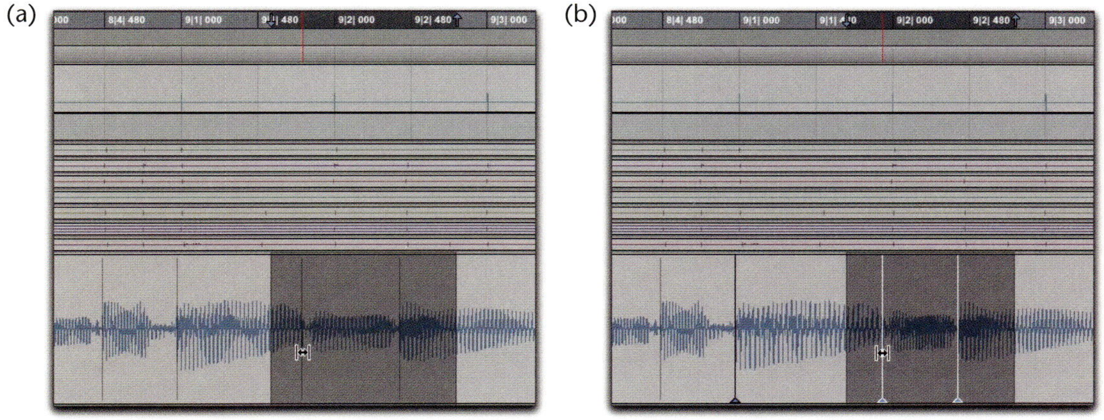

Figure 4.10 (a, b) Warping the Bass Guitar.

Roll forward in the song tightening up the transients of the bass against either the grid or the transients of the kick and snare drums. Since we've already pocketed the drums against the Pro Tools tempo, it's easier to move the bass notes against the grid. However, if the track had not been performed to the Pro Tools click track, you would just want to set up your bass guitar between your Kick and snare tracks, and pocket its transients against them.

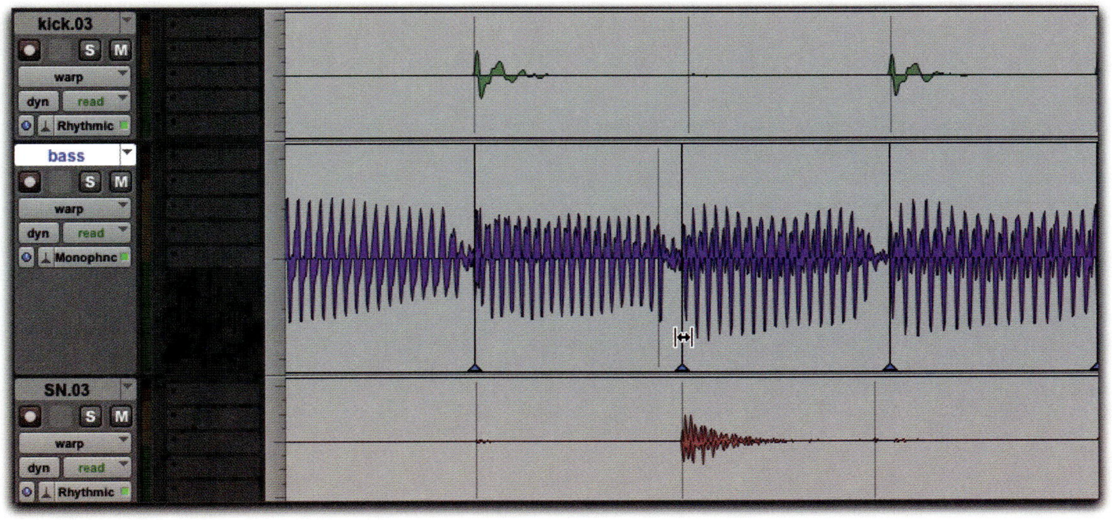

Figure 4.11 The Kick and Snare Can Give a Great Visual Reference to Pocket against When Your Song Wasn't Recorded to the Pro Tools Click Track.

Give it a play and listen. If you made a playlist of the unedited bass guitar track, you should be starting to feel the difference as the low-end punch of your bass guitar begins to sync with the already-in-time punch of the drum tracks. Nice. The beauty of Elastic Audio on something like a bass is that you can achieve significantly better results with much less effort than manually pocketing the entire track. I know, it almost feels like cheating.

Warping to Preserve Bass Attack

Now as you're pocketing, check the note located around bar 25. This is a common problem we have to account for when using EA on instrument tracks … incorrect Analysis markers. In this case, Pro Tools has placed a transient marker both at the front of the note (which is correct) but also just a few milliseconds later during the sustain of the note.

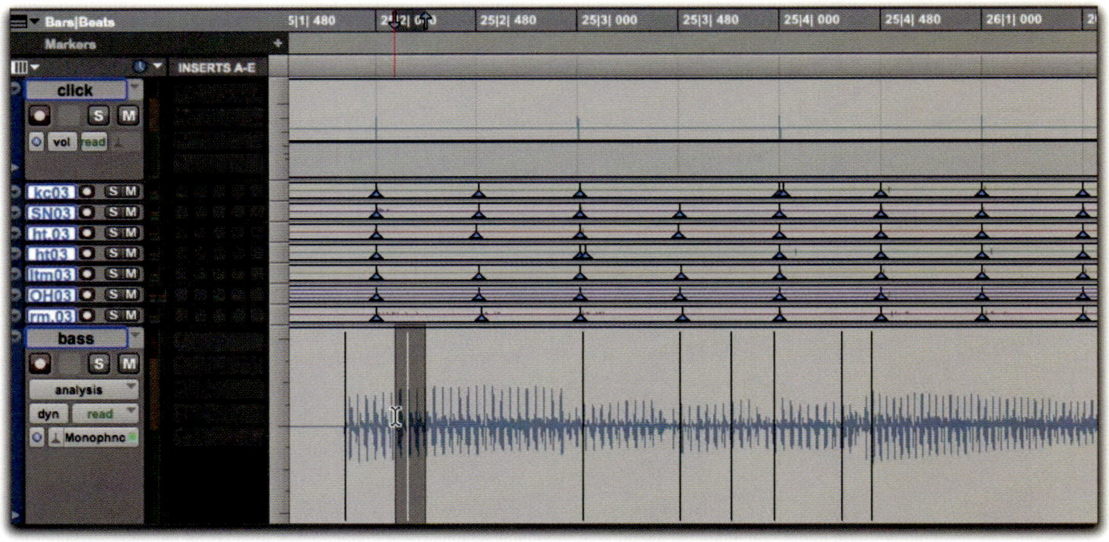

Figure 4.12 An Incorrect Analysis Marker.

Obviously, if you pocket the first Warp marker, it is going to just stretch the front portion of the note, leaving the rest of the sustain in place. That doesn't work as well on a bass guitar sound. Often, we just want to stretch the entire note to minimize glitchy sounds or other audible artifacts.

Our two options are to either move an incorrect Warp marker or delete it. In this case, let's delete the second Warp marker. Switch over to the Analysis view (use the shortcut Ctrl+Cmd+Left and Right Arrow to switch back and forth between Analysis and Warp modes on the selected bass track), select the second Warp marker, and hit the delete key.

Now, holding Shift key, select the Warp marker at the beginning of the bass transient and drag it into the pocket.

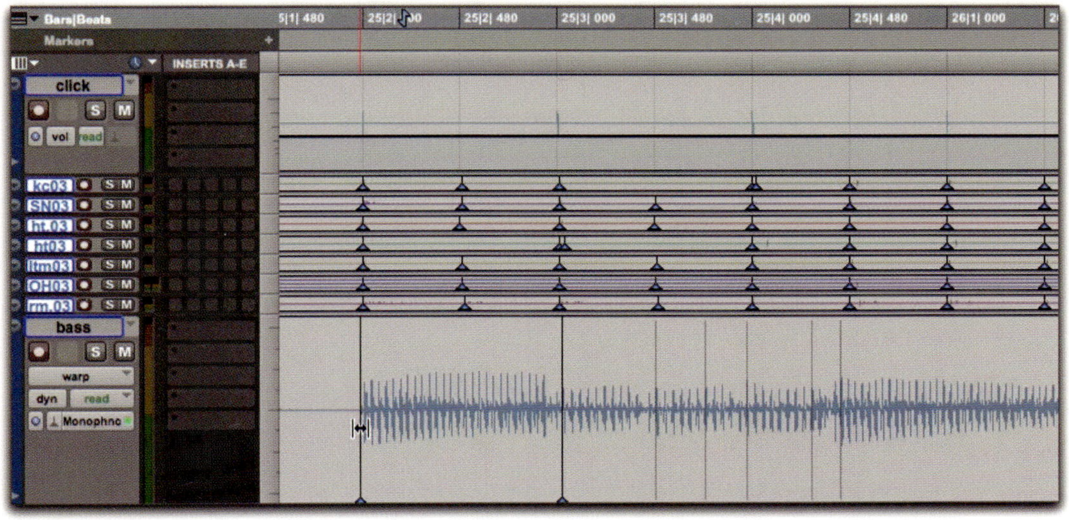

Figure 4.13 Pocketing the Front of the Bass Transient.

I can tell you're getting the hang of this already. Use your ears as you move forward to the other bass notes. If it feels too early or late, go ahead and adjust. Jump ahead to bar 26|3 and move the early note into time.

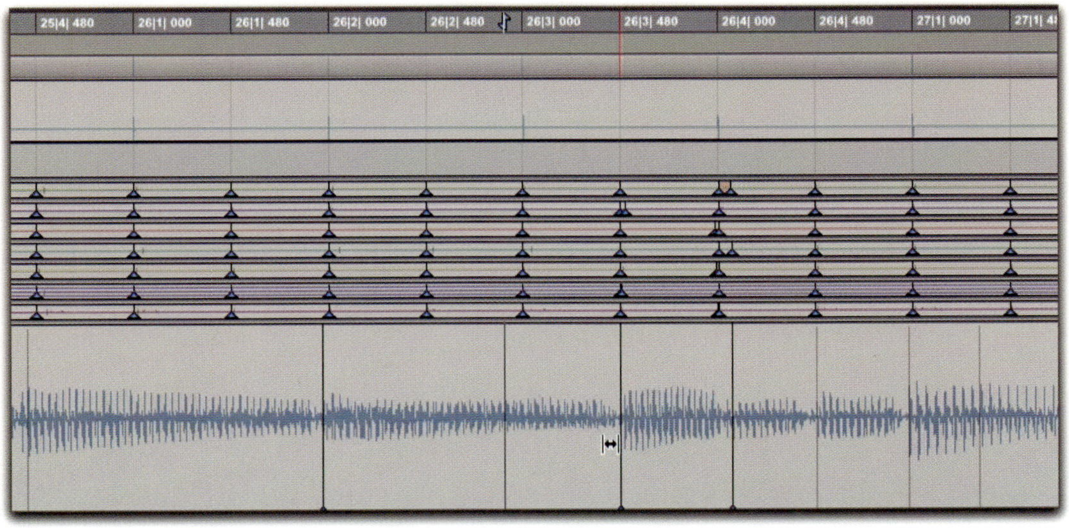

Figure 4.14 Don't Forget to Hold the Shift Key Down to Just Move the Selected Notes.

To move quickly, switch into grid mode, hold down Shift key, and manually drag them to the correct bar and beat.

One of the cool features that Pro Tools 9 brings down from HD is the Continuous Auto Scroll option. This feature will keep the Timeline in the center of the screen, and have the audio scroll past underneath it.

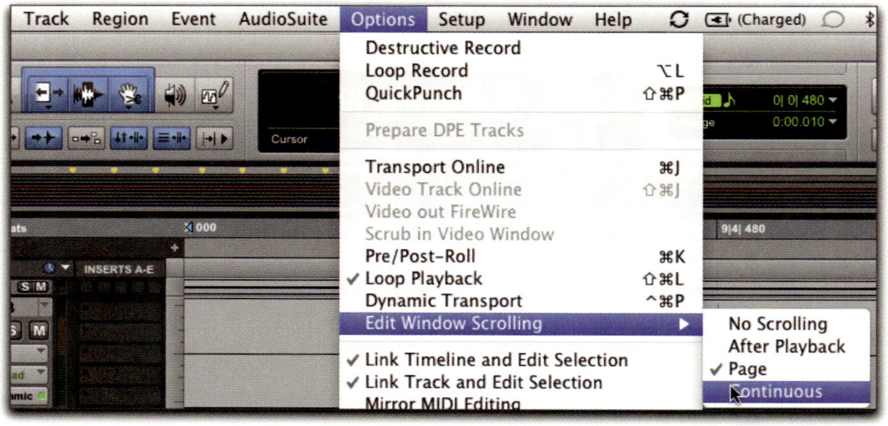

Figure 4.15 Continuous Scrolling Comes to Pro Tools.

Unfortunately, if you try to watch and listen for timing errors with Continuous Scroll turned on at a close zoom level and full tempo, it may give you a headache to watch the waveforms fly by. So, combine the new Continuous Scroll function with the Play at Half-Speed technique by pressing Shift+Spacebar. Now, the waveforms will move by more slowly, allowing you to both listen (albeit at half pitch) and see the notes that are early, late, and in need of correction.

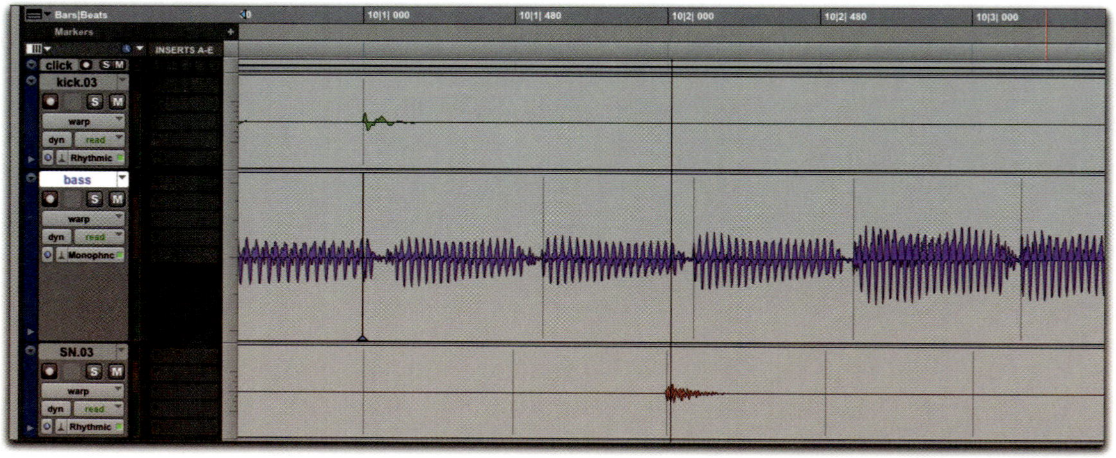

Figure 4.16 This Lateness of the Bass Becomes Even More Visible When Viewed at Half-speed and with Continuous Scroll Turned On.

Around bar 43, you will hear another note with a Warp marker too close to the latter note. Flip to Analysis view (Mac: Ctrl+Cmd+Left arrow, Win: Start+Ctrl+Left Arrow) and let's delete the Warp marker by simply holding the Alt button, which will turn your cursor into a pointing finger with a minus sign beside it. Click on the errant Warp marker and it will disappear. Switch back to Warp mode to grab the Warp marker of the late note just after 43|1 and Shift+drag it into time.

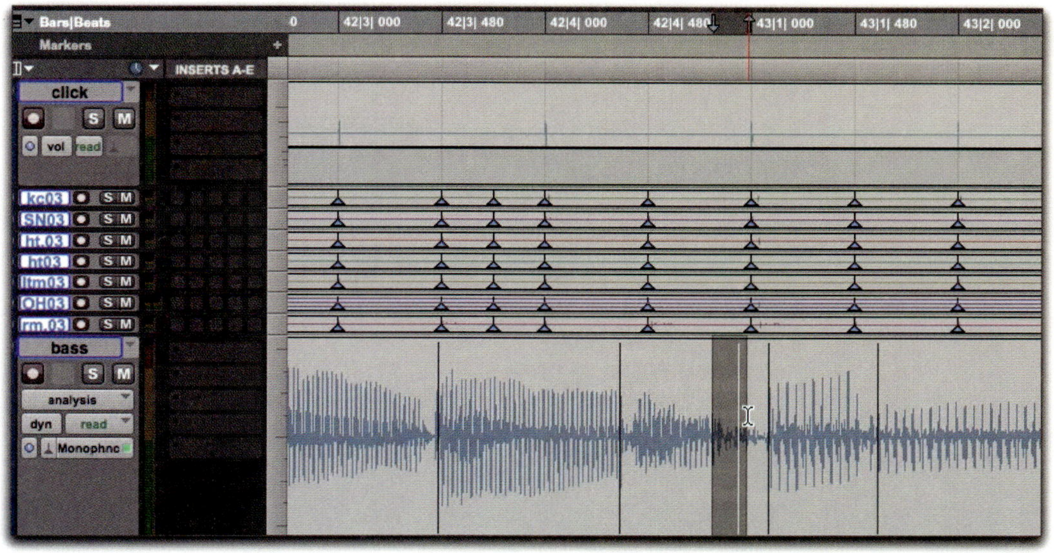

Figure 4.17 Select and Delete, or Alt+click on the Bad Analysis Line to Remove It.

As always, after every edit, listen to a few bars of your track to make sure you're not accidentally time stretching your entire track out of sync. As you get comfortable with the process, there's less danger of this. Still, if you have any doubts, make sure to create a new playlist to edit.

Now what to do if the Warp marker is close, but not right at the transient of the note? At bar 45|2, we have just such a problem.

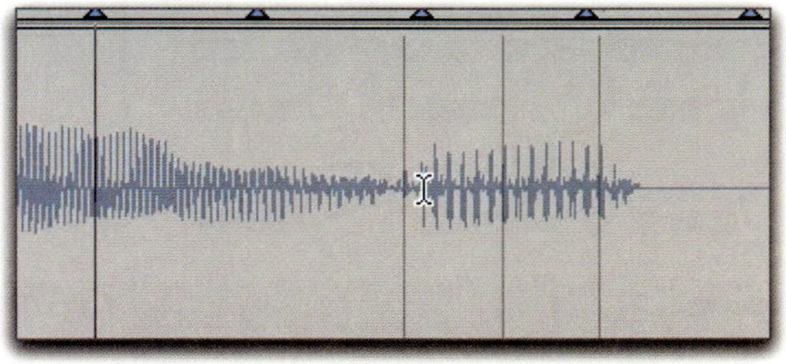

Figure 4.18 The Warp Marker Is Significantly Earlier Than the Beginning of the Note.

The solution is simple. By switching back to Analysis view, we can manipulate where Pro Tools believes the Warp markers should start. Simply drag the Warp marker over to the actual transient of the bass guitar and delete the bad Warp marker in the middle of the note. Now, simply stretch your note into place and you're ready to move on.

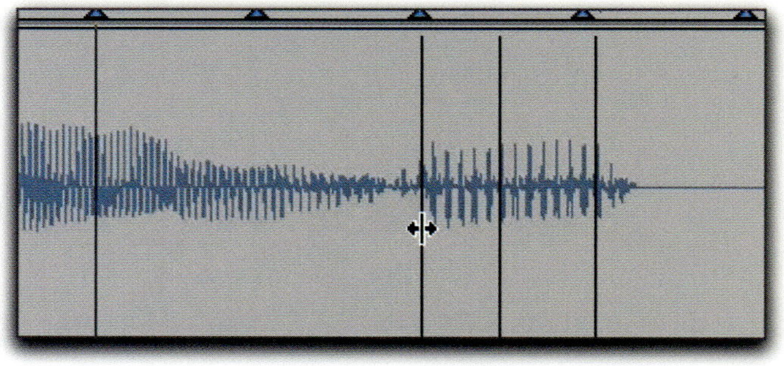

Figure 4.19 Moving the Early Warp Marker to the Correct Start of the Note.

While we're on the topic, occasionally Pro Tools will neglect to put an analysis marker at the beginning of a note (that's why we always trust our ears over software). So, when you ever need to add a Warp marker that has not been detected, simply hold down the Control key in Warp mode and click wherever the missing marker should be.

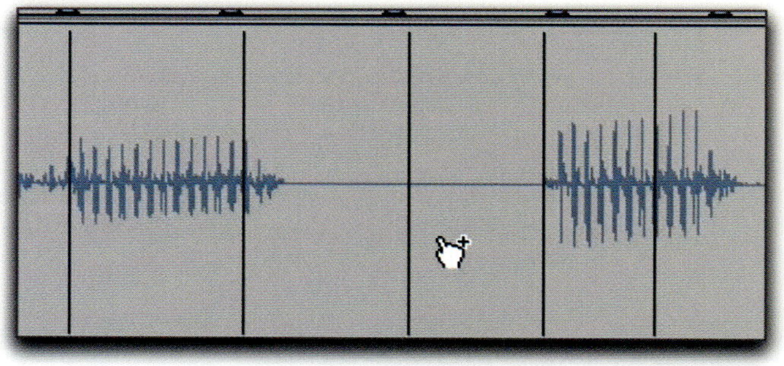

Figure 4.20 The Pointer with a Plus or Minus Will Add or Delete Warp Markers.

Pocketing Bass against Guitars

At about bar 54, you will notice a note that feels like it is out of time, but it's hard to tell when you are listening to just the drums. To rectify this, let's bring in the acoustic and one of the electric guitars so that we can have another point of reference while pocketing the bass.

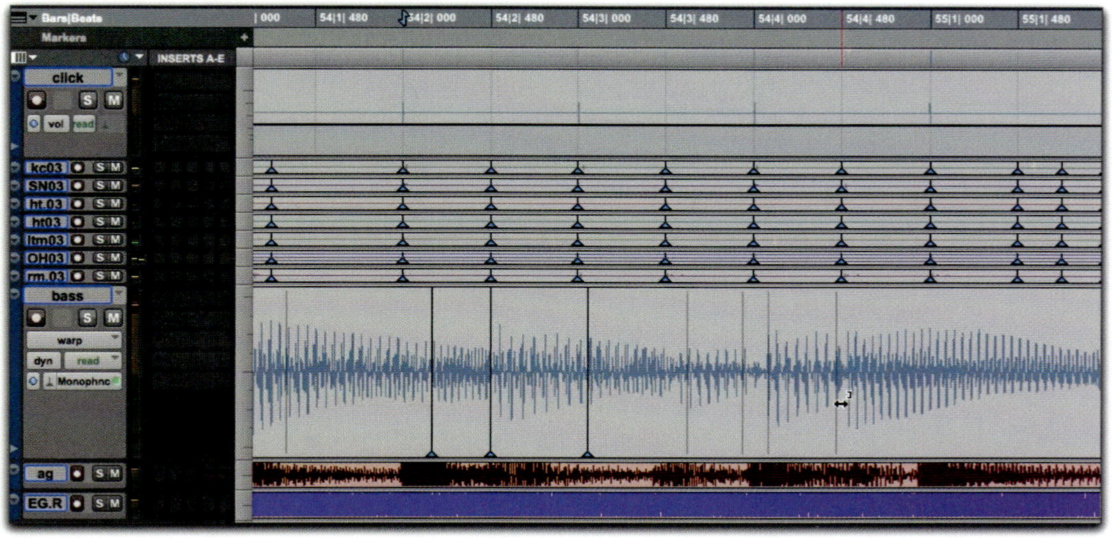

Figure 4.21 The Bass Came in a Full Beat Earlier Than the Rest of the Band.

Sure enough, listening to the guitars shows us that the other musicians waited until 55|1 to hit that particular downbeat. So, while it may be a bit of a stretch, move the transient of the bass all the way over to 55|1. Thank God for good time expansion algorithms.

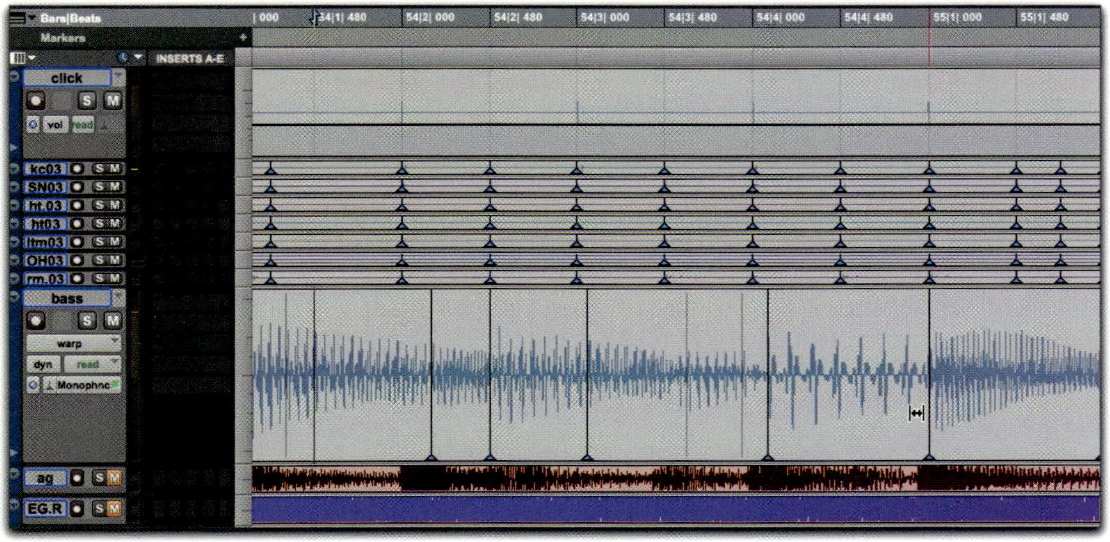

Figure 4.22 Wow. That's a Stretch.

Bass Pocketing Judgment Calls

Now as you listen to your new edit, we're faced with a common mixer's conundrum. While yes, the rest of the band hit the note at 55|1, when we listen back, it almost sounds late. So, the performer intends to play the note that way, or was it simply a matter of being out of sync with the other players? After listening, I think I'm going to move it back to 54|4, and if it continues to bother us, we will fix it. As a precaution, let's drop in a Memory Location by hitting the Enter key on our keyboard numeric keypad, just in case. Name it with something memorable, so we can come back later and find our potential edit point.

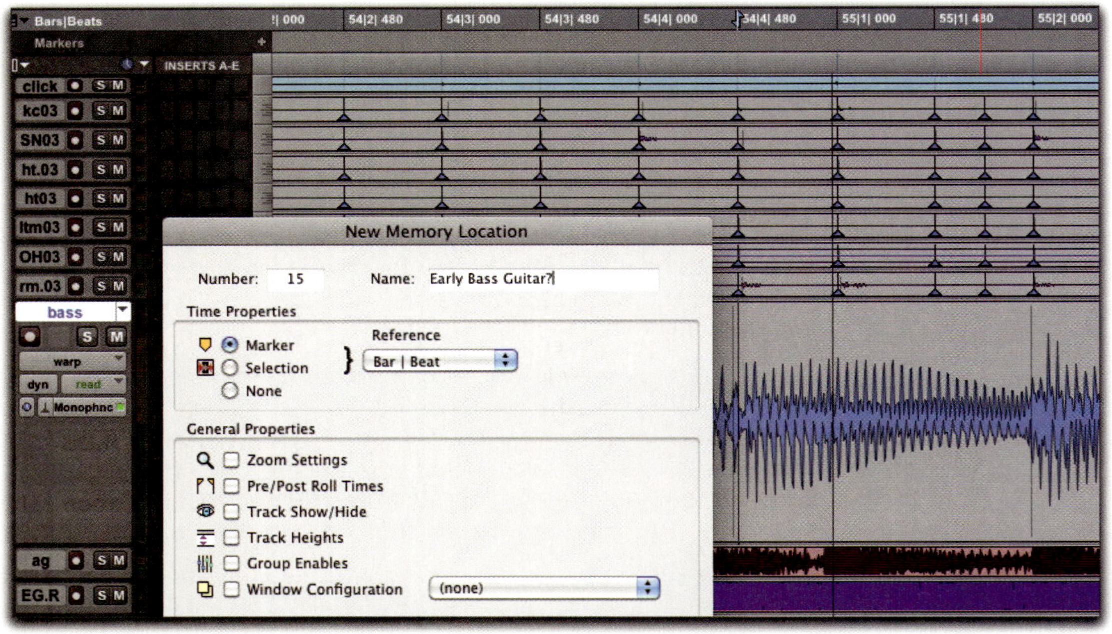

Figure 4.23 When in Doubt, Drop in a Memory Location of Things to Check Later.

At about bar 64, you will hear yet another note whose Warp marker actually appears noticeably before the actual transient. Rather than attempting to move the early Warp marker to actually be correctly on the transient, since we're manually pocketing this, the fastest way is to switch back into Slip mode and simply pull the transient of the bass guitar into line with the drums, which is easy as can be.

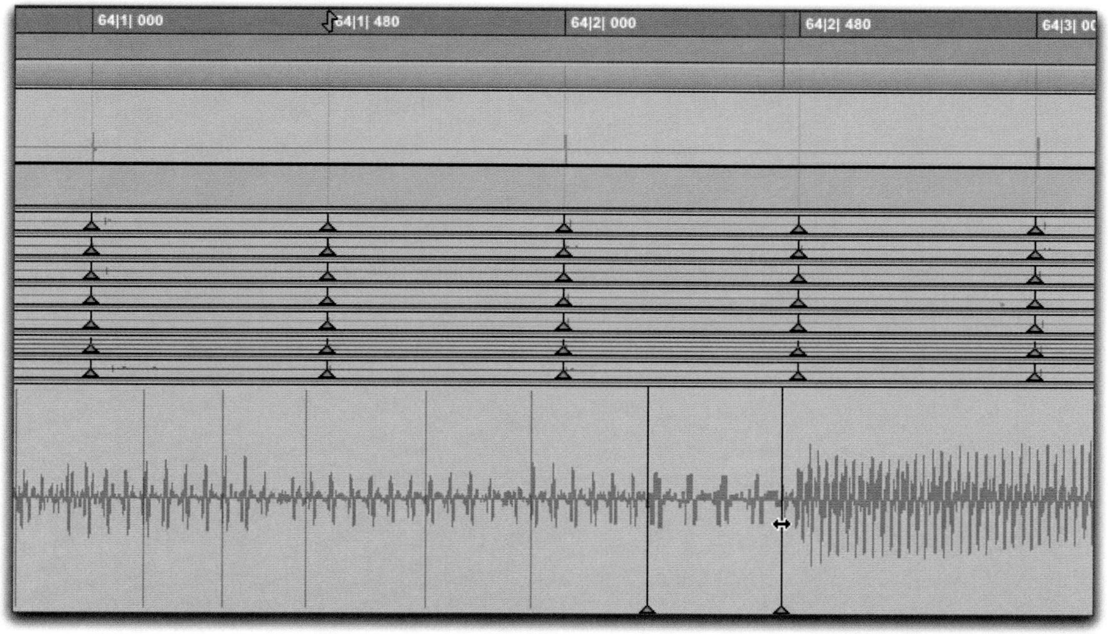

Figure 4.24 Sometimes It's Faster to Just Visually Align the Transients Than to Fix Every Warp Marker.

Go ahead and continue on editing the remainder of the bass guitar track. You will hear some noticeably late notes around bar 75 that can be fixed with our simple Shift+drag method to align the transient of the bass with the closest drum note. Of course depending on how tight or loose you want the song to feel, you can pocket every single note as we commonly do for most major recordings. Even with A-list players, it is not uncommon to spend 5 to 6 hours pocketing a track to produce the feel of world-class players on their best day.

Alternatively, if you are just need to do a quick tightening on a low-budget project and you've sufficiently developed these skills, you can probably work through a reasonably tight track in about an hour. Your mileage may vary.

Other obvious timing tweaks required on the bass guitar can be found around bar 93, bar 108 to the break, and bar 120 or so to the end. Take the next 15 to 20 minutes and finish out the bass guitar through the end of the song. I think you'll agree, Elastic Audio can be quite a time saver versus manual pocketing, and it doesn't sound half bad either.

In This Chapter

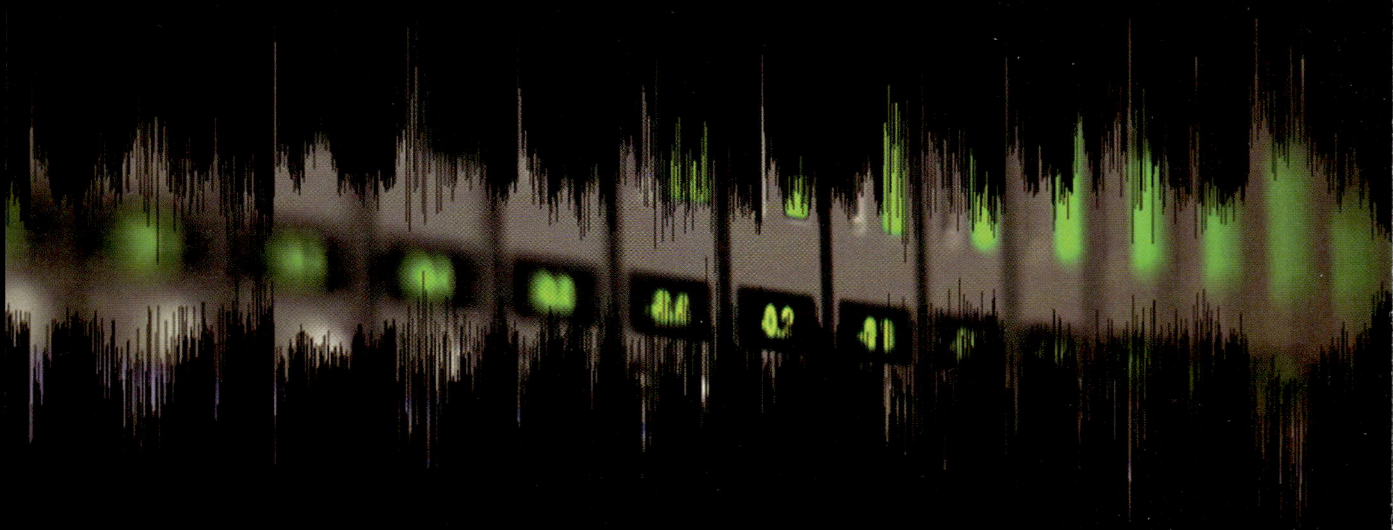

5

Guitar Mixing

Roughing in Guitars

All right, now that our drums and bass have been pocketed, edited, tweaked, and tucked, let's take a look at roughing in these massive stacks of electric and acoustic guitars. Depending on how much time you spend editing the drums and bass and to what degree you have pocketed them, you should be ready to start adding the guitars to the mix.

As a starting tip, especially when you are working on a song that you were not the tracking engineer for, take a look through the various guitars and you will quickly see several that only play once or twice in the entire song.

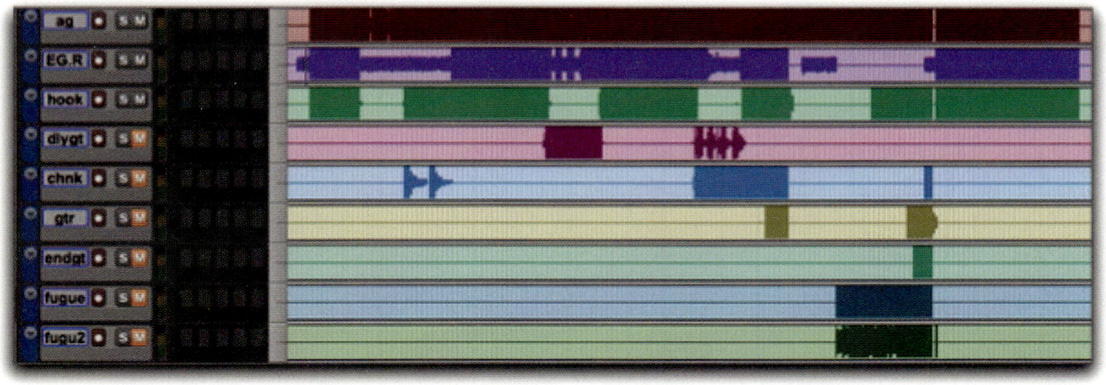

Figure 5.1

In most cases, those will only be special parts that we will look at more closely after we have built up a solid foundation for our track. So for now, mute the fugue guitars, and let's look at the one acoustic and two electric guitars that play all the way through, and then we will move on to adding the special guitar parts.

© 2011 Nathan Adam. Published by Elsevier, Inc. All rights reserved.

To start we have the guitars panned hard left and right, which is a good starting point for when you have two guitars performing similar parts. Of course, as the mix develops, we will be tweaking this further.

Creating a Master

Before going any further, let's create a master fader for the whole session. Used the Shift+Cmd+N (Win: Shift+Ctrl+N) shortcut to bring up your new tracks dialog, then holding the Command key down, use the right and left arrow keys to select between stereo or mono audio tracks, then the up and down arrow keys to toggle between track types. Use the Command (Win: Ctrl) and down arrow key until you see the master fader track type and hit the Enter key on the lower right corner of your keyboard to create it.

Figure 5.2
Creating a Master Fader.

Of course you could have done all of that with your mouse, but that would be slow as molasses, and not the Mixcoach way. Drag the master fader all the way to the left-hand side of your session.

Figure 5.3
Moving the Master Fader.

Mono-Fy Your Mix

Now's as good a time as any to show you one of my secrets to building better mixes; develop your mix (and initial levels) *in mono.*

"Madness!", You Say. No One Listens to Anything in Mono Anymore...Right?

Mono is the worst case scenario you can listen to a mix in. You want to make sure your mix sounds the best in the worst possible conditions. Whether that is playing over your friends crappy car stereo or out the 70-V mono sound system in your local shopping mall or restaurant, if you think about it, as soon as you are not sitting between the two speakers in the sweet spot, any time you hear music it has effectively been summoned to mono. Therefore, if it sounds as good as it can possibly sound, in the worst possible conditions, then it can only sound better on an average to great sound system...or even when fired directly into someone's ear canal over their iPod earbuds. You're welcome.

There are a variety of ways to sum your mix to mono in the control room. The simplest way inside of Pro Tools is to put the AIR Stereo Width plug-in across your master fader, which will allow you to set a width of 0%. This will effectively sum all of your panning and tracks to mono, which will allow you to hear any significant level or face problems in your mix early on. Then, when you have developed your mix to a greater degree, you will be able to quickly bypass this plug-in and hear the full stereo glory of your tracks.

Figure 5.4 Using the AIR Stereo Width Plug-in to Mono-fy Your Mix.

Now before we start to edit these guitars, let's build a balance between the two guitars in mono. As you start to play, you will hear the electric track has a nice big "distorted guitar" sound, but it has too much low end.

The Filtering Secret

Wait, what? No guitar has too much low end … right? Lesson number one in mixing is that every sound can't be full range. If you think about it, that is, all excess energy that was captured by your microphone (there may be air-conditioning noise, rumble of the microphone stand, breathing, and so on) does not need to be added to your mix. It will only conflict with the kick and bass guitar while competing for room in the low end, making your mix wubby and woofy.

This is one of the number one secrets of making a mix with tight bass: filter out the low garbage from any track that doesn't require it.

To accomplish this, lets put on a seven-band EQ and to use the low-cut filter to roll off some lows.

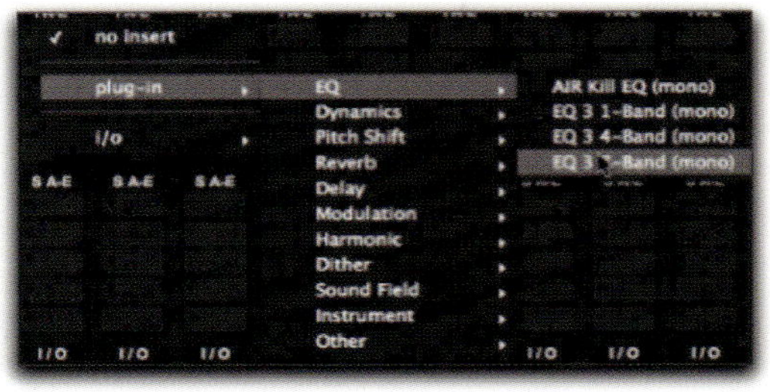

Figure 5.5 Drop That EQ On.

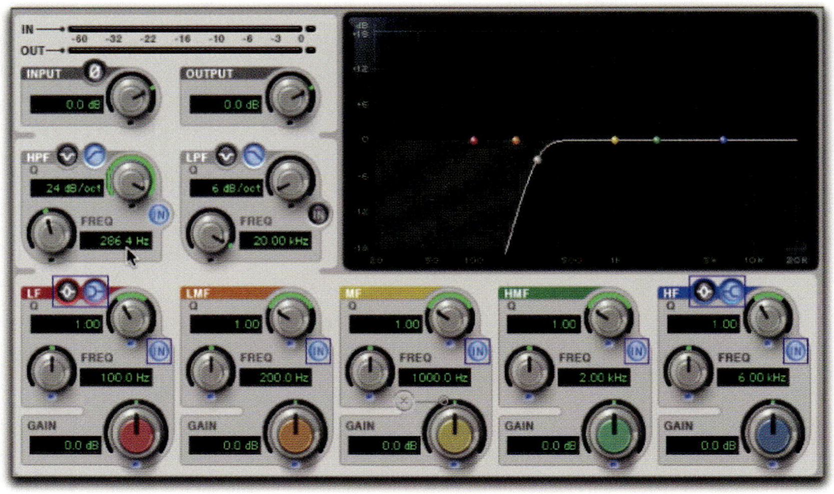

Figure 5.6 Engage Your Low Cut or HPF (High-Pass Filter) and Roll the Frequency Knob Up and Down to Hear the Effect of the Lows Being Filtered.

Basically, what the HPF does is allowing everything *above* the *frequency* knob to pass through, while frequencies below are filtered out. The other knob on your filter adjusts the slope or the rate at which the filter cuts off eliminated frequencies. Start to sweep the frequency knob lower and higher and you will hear the low frequencies being cut out of your electric guitar.

So how much should you roll off? Here's my standard approach for filtering out excess low end on an instrument: Start with a medium slope of 12 to 18 dB/octave. Switch the low-cut (also called a "high pass," because it lets the "high's" pass … get it?) filter on, and start to sweep the frequency knob up until you hear the low frequencies in the guitar start to go away. With many instruments, you will hear that you can cut off a great deal of low frequency before it even starts to affect them audibly.

Figure 5.7 Start Rolling the Frequency Knob Up Until You Hear the Unnecessary Lows Go Away.

With this guitar, we get almost all the way up to 127 Hz with an 18 dB/octave slope. Now, if you bypass the plug-in back and forth so that you can compare the pre- and postfiltered sound, you will notice some of the low end now missing from the electric guitar. That's okay. We are building space in the mix for our other low-end instruments to reside.

Figure 5.8 Bypassing the Plug-in Let's You Compare the Uneffected Sound to the Sound with the HPF.

It's a common complaint among amateur mixers to note that their drums sound huge and aggressive when soloed but bland and mushy in the mix. Much of the time, this is because of the excess low frequencies piled on by all of the other audio tracks. It's literally like putting too many ingredients into a cake mix. A cake can only accommodate so many eggs, so much flour, or so much sugar. Too much of any one ingredient and the cake won't work. Man, now I'm hungry. Where was I?

Initial Guitar Levels

Now that we have filtered our electric guitar track, solo the Hook guitar track.

Let's take the expedient route for a minute, and copy the EQ from our electric guitar to the Hook guitar by simply holding Alt and dragging the plug-in over. This makes a duplicate of the plug-in, complete with settings, on the Hook track. That was fast.

(a) (b)

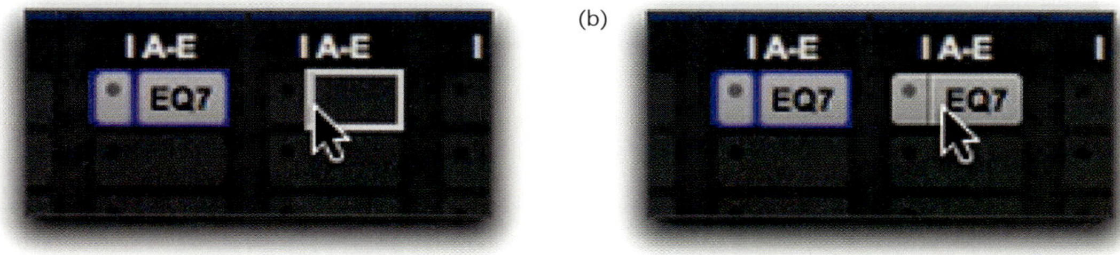

Figure 5.9 (a, b) Copying the EQ Plug-in by Alt+Dragging.

Now, repeat the process and sweep your filter frequency up and down until you find a low-frequency cutoff that eliminates all the low end you *can't* hear, and maybe just a touch of low end that you can hear. In this case, since we know it was the same guitar player doing a second pass, we can keep the HPF at the same frequency and tweak later if need be.

Now that I have the guitars taking up less room in the frequency spectrum, let's blend their faders into a good balanced level with the drums and bass. I don't know about yours, but my faders ended up somewhere between –14 and –17 dB as a good starting point.

Figure 5.10 My Blend of Guitar Levels.

Now that they are placed in the mix, let's take a look at the acoustic guitar. Acoustic guitars are notorious for taking up lots of space in the frequency spectrum. So in an effort to save time, let's copy the EQ from the electric guitars and start to sweep our high-pass filter around again. Up around 200 Hz, our acoustic starts to get too thin, which we don't necessarily want. As we move it back toward 100 Hz, we start to filter some extreme lows without sacrificing the tone of the acoustic. This has created a tremendous amount of space in our mix for the kick and bass to keep things punchy. That's rock and roll.

(a)

(b)

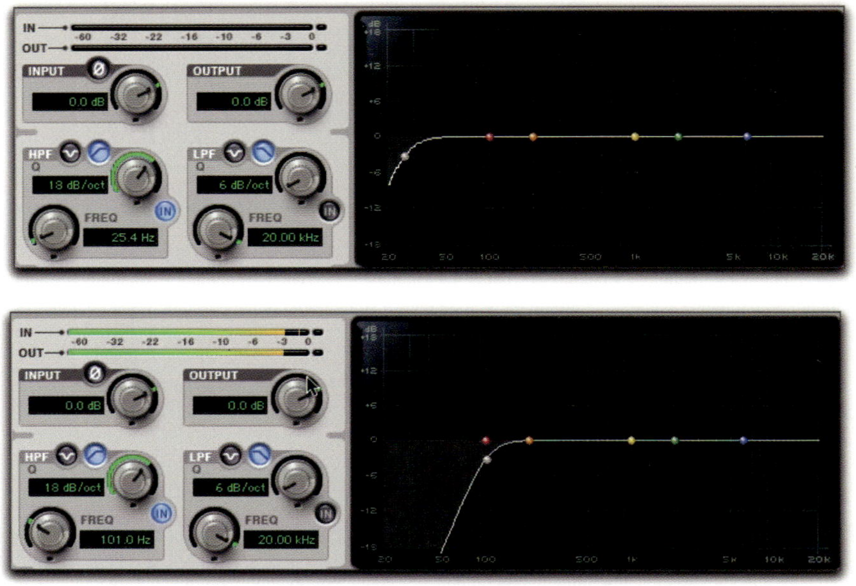

Figure 5.11 (a) Start Here; (b) End Up Around 100 Hz.

Now, start playing through the intro on mute in your electric guitars and blend your acoustic guitar track in until it is providing a solid punchy rhythm. My fader ended up around –14 dB.

Acoustic Guitar Compression

Now that we have three guitars going through the song blended in, let's avoid a common problem. Acoustic guitars in a rock and roll track like this tend to get lost in the mix. If you have a channel strip plug-in, like the SSL or API channel strips, I tend to use them. Using the stock plug-ins, start with the bomb factory BF 76 compressor plug-in to try to help level your acoustic guitar out.

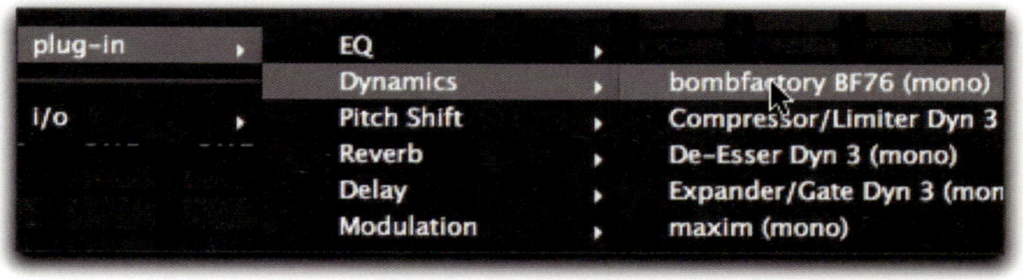

Figure 5.12 The BF76 Is a Great Choice of Compressor Plug-in.

This is a great plug-in modeled after the classic 1176 compressor. Initially, the only settings we will change are to make the attack as long as possible (all the way to the left at 1) and the release as short as possible (all the way to the right at 7). Keep your ratio at 4:1, and then let's adjust our input level by turning it up as the guitar is playing. Start with the compressor plug-in bypassed and the two electric guitars muted.

Figure 5.13 Glorious Compression.

This will allow us to hear the acoustic guitar compressor working more clearly. As you play back the acoustic, switch the BF76 BYPASS button in and out, which will allow you to hear how it's leveling the volume transients of the acoustic.

Now notice what is happening to the acoustic guitar. The compressor reduced the apparent dynamic range, which allowed me to compensate by pushing the fader up a bit. When he strikes the strings really hard (which we can see in the very first strum of the song around bar five) by using a slow ATTACK and a fast RELEASE, we are allowing for transient of the acoustic guitar to make it through the compressor, and only the sustain of the guitar to be compressed. Furthermore, because our ears latch on to the loud transients, we perceived it as getting louder.

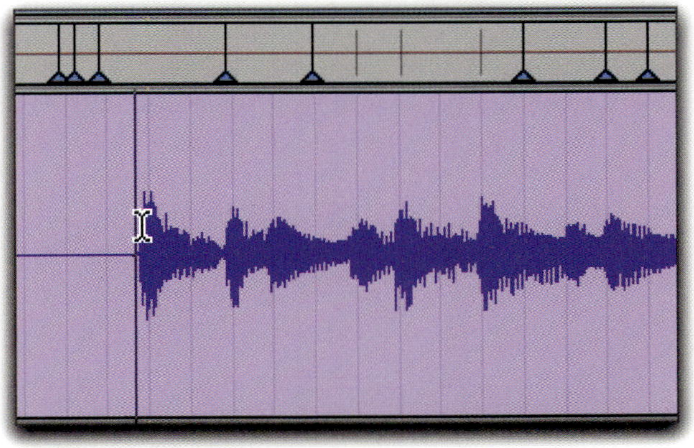

Figure 5.14 Just Look at That Acoustic… Waiting to Get the Smackdown.

ABing Compressors

Compressors are especially useful on a guitar, especially if you have classic compressors like an LA2A. However, since Avid isn't necessarily including an LA2A plug-in (hint, hint…is anyone listening?) in the Digirack plugs, let's do a little AB testing by adding the Compressor/Limiter 3 after the BF 76.

(a) (b)

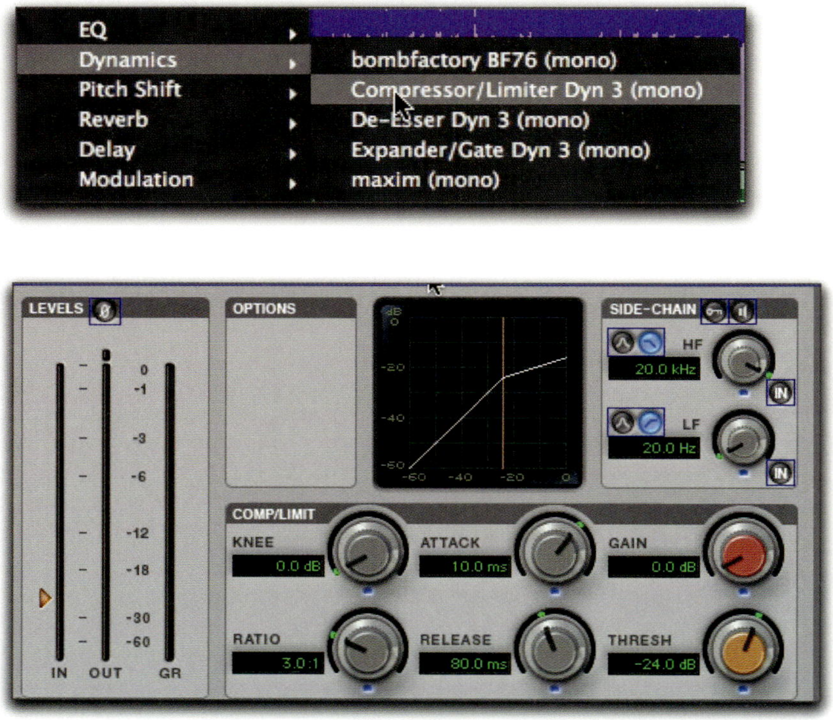

Figure 5.15 (a, b) The Stock Digirack Compressor Plug-in.

Now, for a true A/B, we should attempt to set it up with the exact same settings. However, you'll note by their wildly different interfaces that this is easier said than done. So, rather than a perfect A/B, let's pull a sneaky trick out of the pro-mixer's toolset and use … wait for it … a preset.

I know, it offends your very sensibilities to even hear it, but after mixing almost 1000 records, I can openly admit that I don't mind listening to the presets included with any plug-in, especially as a starting point.

Typically, especially if you are new to the recording and mixing process, you'll find that some great engineers have spent serious time giving you starting points for your compression, EQ, and effect settings in these plug-ins. Although these will obviously not work for every sound and situation, you will find that they are a great jumping off point for new and old engineers. As always, trust your ears.

So, start with the Acoustic Guitar preset on the new compressor, and bypass the BF76 plug-in by command clicking its insert. Hit play and listen to the new compressor setting.

(a)

(b)

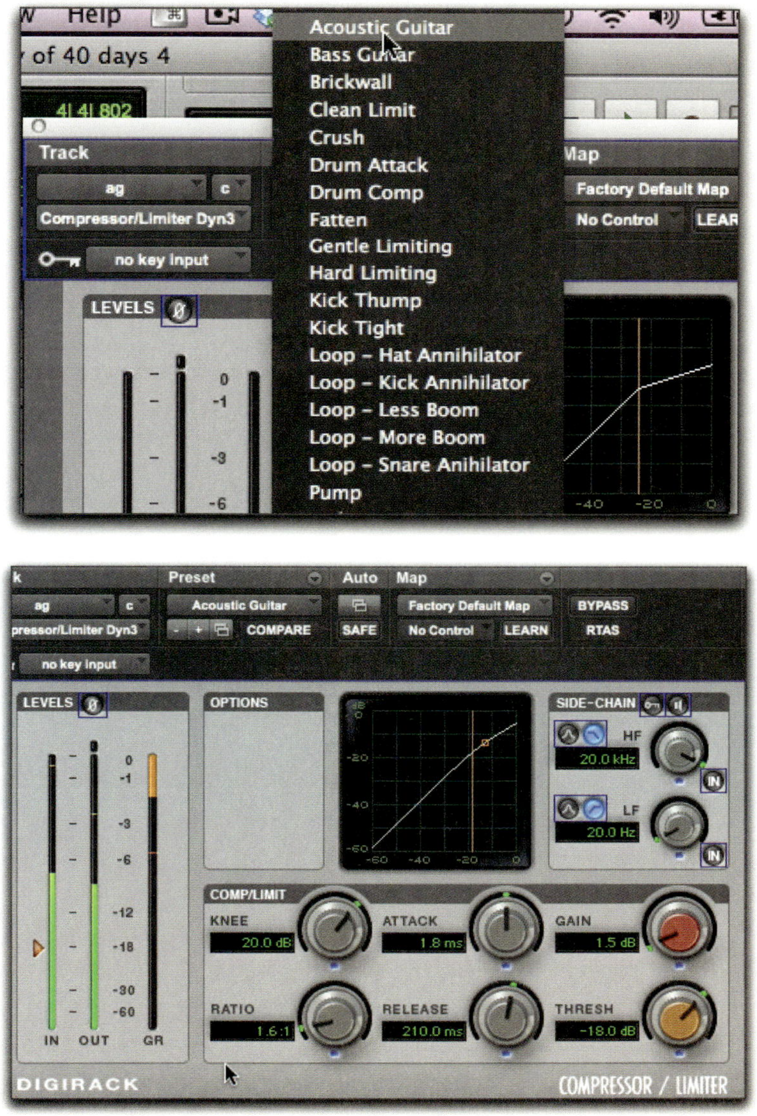

Figure 5.16
(a, b) Presets Are a Mystery to the Uninformed.

If you play through the first chorus, bypassing, and un-bypassing the new compressor, you should hear the sense of urgency that a compressor has to impart on an instrument. When you bypass the compressor, it will almost sound like he has gone from playing angry, to playing happy. Pretty wild huh?

(a)

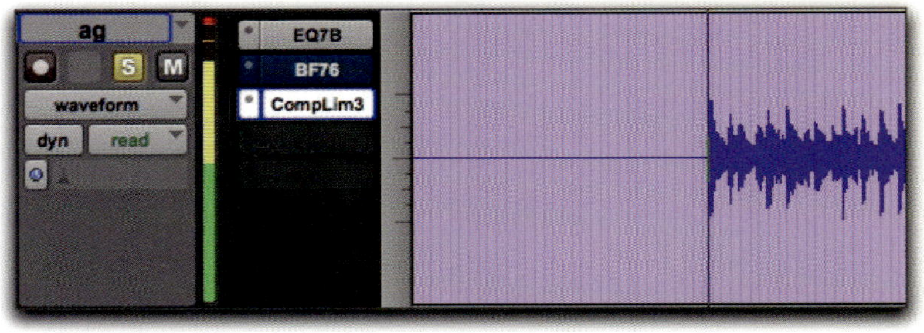

Figure 5.17
(a) Bypassing and Soloing Are Two Great Ways to Develop Your Ear for the Subtler Niceties of Mixing.

Believe it or not, I actually like the urgency of the second compressor as opposed to the smoothness of the BF76 for this particular track. Of course, as we move through the mix, we may find that other tweaks need to be made. As you will learn about the process of mixing, a decision is never final, just waiting to be recalled. Go ahead and remove the BF76 and un-bypass your Compressor/Limiter 3.

Toggling between Presets

Finally, we get to some real rock and roll. Unmute your electric guitar tracks and let's adjust their blend relative to the new compressed acoustic. The first thing I hear is that the Hook guitar, because it is playing an urgent part, could stand some aggressive compression as well. Drop the same Comp3 plug-in on the Hook track after the EQ plug-in, and go through the various presets while really listening to see how each one affects the electric guitar. A quick way of accomplishing this task is to use the plus and minus buttons located beneath the preset list to the left of the COMPARE button.

(b)

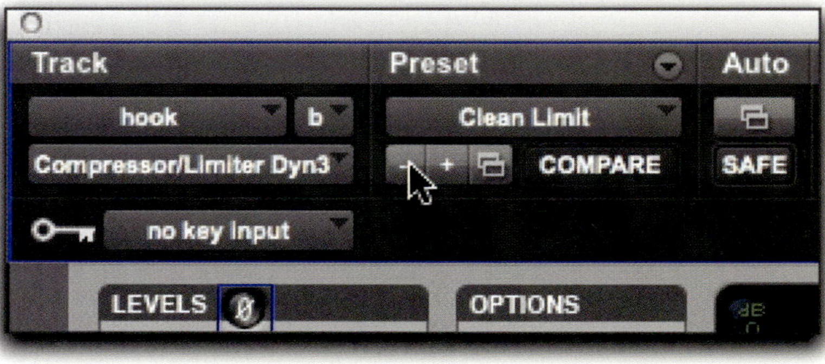

Figure 5.17
(b) I Love the Fast Way to Toggle between Presets.

As you toggle through them, you should hear some really dramatic differences in the Hook guitar. Some presets will make it very aggressive and in your face, whereas others will lay the performance back like a smooth-jazz metal head (those exist, right?). By quickly toggling between presets, it allows your ears to focus more on the sound. Assuming you have done that, let's start with a Clean Limit preset. I like the way it makes the Hook guitar sound both louder and cleaner at the same time.

Now, as a reminder, if you feel like the track sounds a little lackluster, remember that you are mixing in mono. And although mono certainly sounds worse, it *is* allowing us to better balance our levels while hearing any phase problems with the current tracks. If we start off mixing in stereo, there is a tendency to just throw the tracks wide right and left, and end up with a mix that sounds very unbalanced in mono. If you need a little encouragement, just bypass the Air Stereo Width plug-in on your master track for a moment to remind yourself you are an awesome mixer in the making.

Figure 5.18 Give Yourself a Momentary Break from Mono-Land. You'll Be Surprised at How Much Better You Mix When You Force Yourself to Learn in Mono.

Mixing Auxillary Guitars

I'm feeling pretty good about the main guitars and their blend, so let's take a look at the other sporadic guitar tracks. Soloing the delay guitar track, below the Hook guitar, and jumping to its performance at bar 46, I'm hearing the same low-end problem we heard on the other guitars.

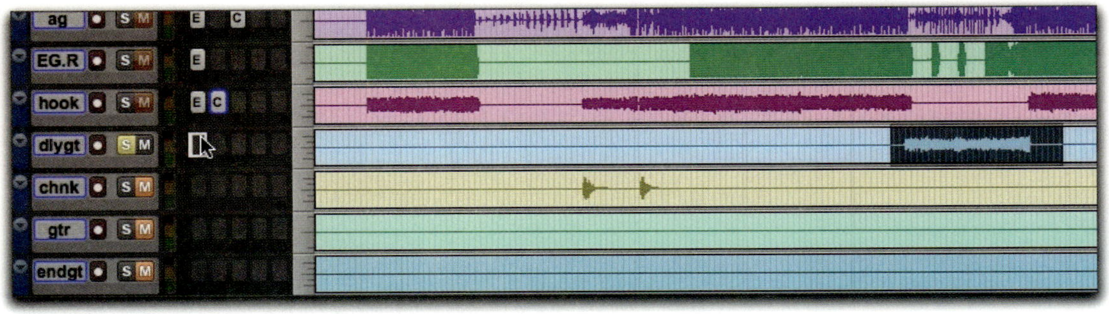

Figure 5.19 Copy the EQ and Compressor by Alt+Dragging Them from the Hook Guitar Track.

Since I know this is the same player, as many of your own tracks will be, let's copy both the EQ and compressor from the Hook guitar as a starting point. All great mixers are about developing a workflow, and finding a great starting point quickly is one of the key methods to keeping your creative juices flowing. This is why top mixers like Chris Lord use a known quantity of processors, console, and track order on every major mix they do. It allows them to focus on the details, not on finding a new compressor for every track.

Once you have tweaked, solo and listen to it in the track. For the rough mixing process, I like where it's placed, so let's move on to the Chunky guitar.

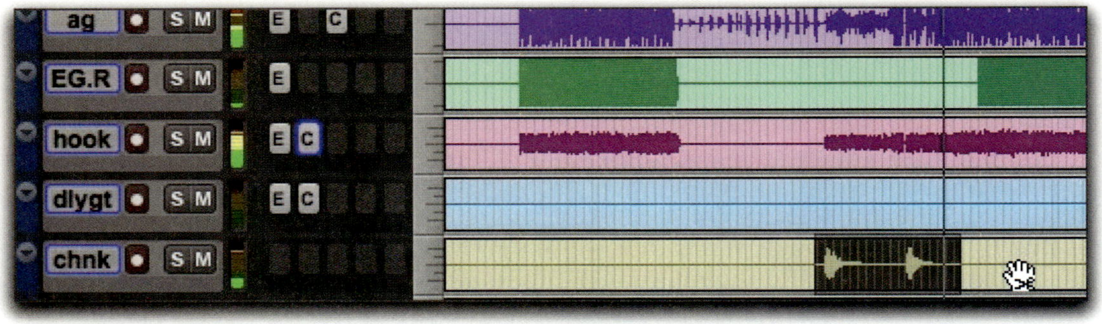

Figure 5.20 Why Does a Guitar Named "Chunky" Start with Two Clean Strums?

The first few strums are fairly clean, whereas when you jump down to the performance at bar 70, you will hear a Chunking performance that could use a little compression. Again, quickly copy your compressor plug-in, and cycle through a few presets to see if any feel better when played in the track. I kind of like what the "Brick Wall" preset does to this guitar, but let's stick with Clean Limit again, and we will tweak later as the track develops.

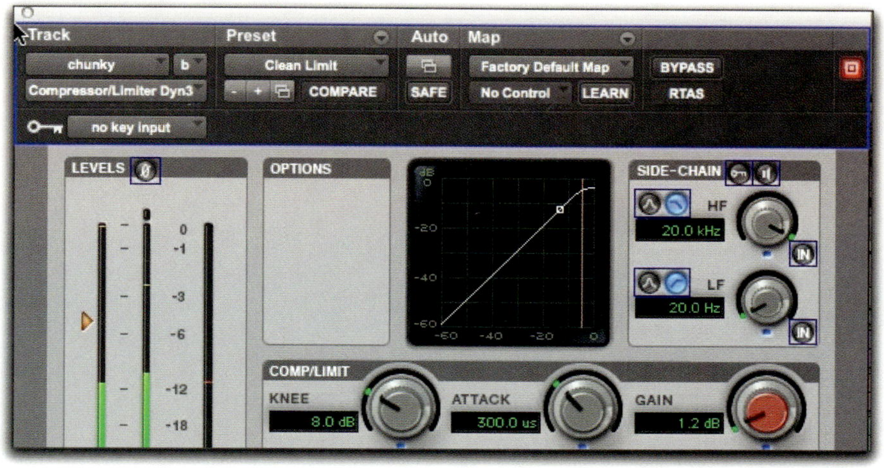

Figure 5.21 Compressing the Chunky Guitar.

Holy cow, yet another guitar track at bar 80 playing a cool, clean lick. Copy the Clean Limit compressor from the Chunky guitar track.

One thing I like to do when putting guitars into a mix is to imagine, "If every one of these guitar tracks was a different player, and all of their mothers came in to listen to the mix, would mom be able to hear her child in the mix?"

Prioritizing Track Secret

While I realize that's kind of a goofy analogy, it will really help to set your perspective as you attempt to fit various parts in. So, while this one little guitar part may seem small and insignificant, we should at least try to give it a fair shot of being heard. Overall, it's a good volume until his loud part at the end, so we will end up writing the automation on that part at the end.

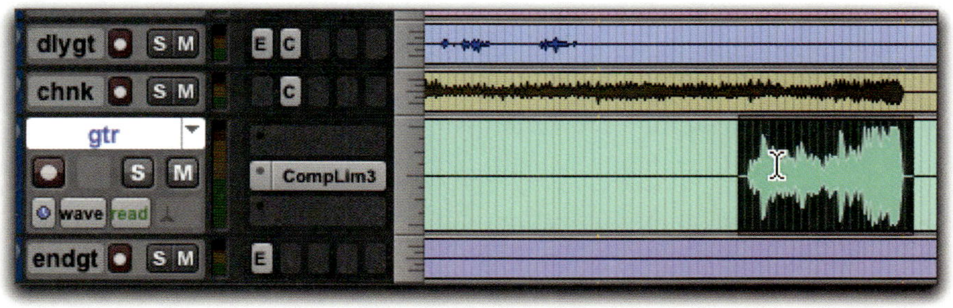

Figure 5.22 Compress Him and Move On.

Finally on the end guitar track, let's copy the same high-pass filter from the earlier electric guitars. Again, we are setting up our rough mix so that no part jumps out and kills our ears initially. This is the first but crucial step of building a great mix.

Figure 5.23 Alt+Drag the EQ from the Delay Guitar Track and Tweak the Filter Frequency for the End Guitar Track.

Fugue Guitars

Now that we have all of our main electric guitars roughed in, it's time to look at these two fugue guitars. Solo them out, pan them wide for a moment, and give them a listen.

Figure 5.24 Go Ahead and Pan the Stereo Fugue Guitars Wide.

EQing Clean Guitars

Do you hear the low end building up in the amps? It's a great sound when you're playing the guitar by yourself, but in a rock and roll track, with the drums and bass, the low end on the fugue guitars is not going to be heard anyways, it's just going to muddy up the track. As a common starting point, Alt+drag your electric guitar EQs and compressors to these tracks as well.

Now that we've got the compressors and EQs on the fugue guitars, the low-end build up on this is even *more* prominent. Lest you think we're always just going to copy a single EQ across multiple guitars (you know you were beginning to wonder). Let's identify the offending frequency and annihilate it.

Believe it or not, the best way to find an offending frequency in a track is to fire up your EQ and try to turn up the bad frequency even more. Then, when you've identified it, we will reverse the gain and cut the nasty sound out.

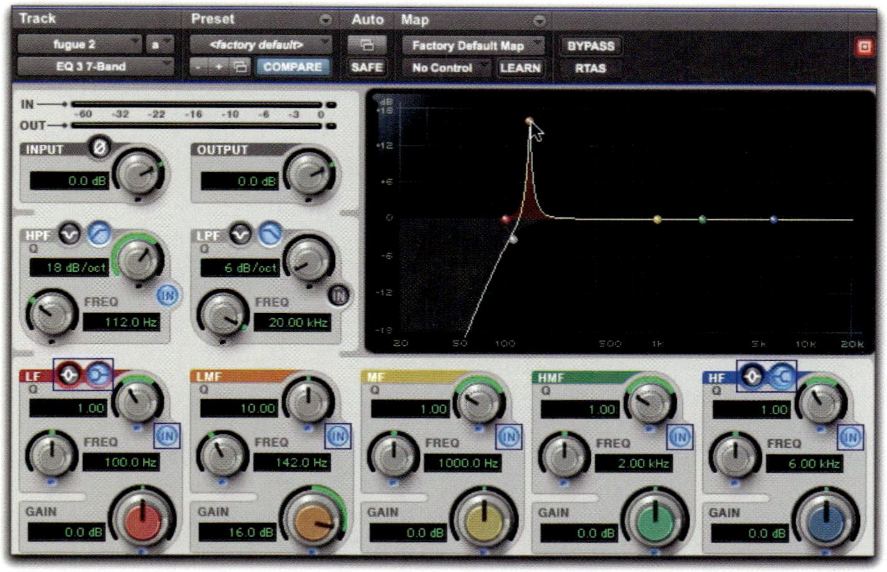

Figure 5.25
Finding the Ugly Low-End Buildup.

Start by setting the orange low–mid frequency EQ to a very narrow *Q* of 10, with a giant boost of around 16 dB. Now start to sweep from the low to the mid range frequencies slowly, listening for the low-frequency junk sound to become even more pronounced.

I'm hearing it centered around 139 Hz. So, let's just reverse the EQ by pulling the Gain knob down to about –3 dB. Nice.

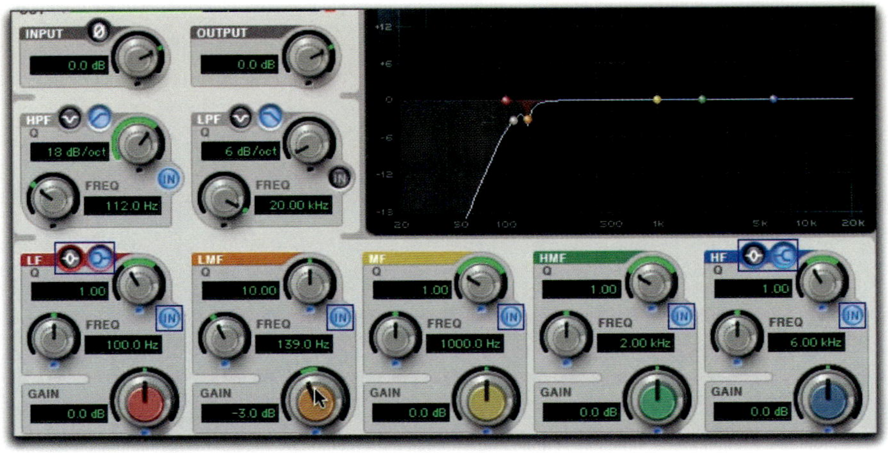

Figure 5.26
A Tight Notch EQ to Pull Out Unnecessary Frequencies Can Really Start to Clean Up a Track.

Very cool. As you keep listening through, it feels to me like the fugue guitar is a little thin for the track that's coming together, so let's EQ it a bit in two ways: (1) Move the HPF cutoff frequency down a bit, restoring some of the fugues lows and (2) Find a nice low frequency on this guitar to boost.

To do this, we simply repeat the previous steps we used to find frequencies to remove, but with a few extra steps on the end.

Start by boosting the LMF EQ with a medium *Q* (~3–4) and gain (11–15 dB). Now sweep the frequency around the low mids listening for the sweet spot frequency that adds a warm bottom end to this particular guitar track.

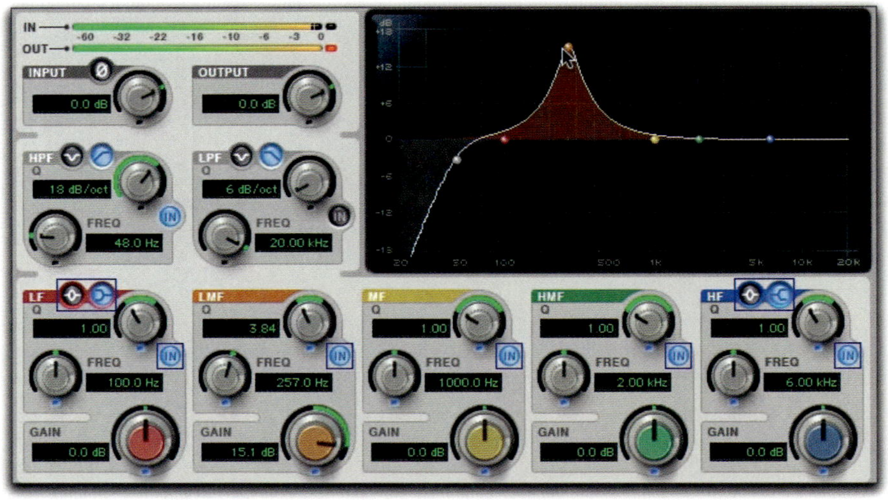

Figure 5.27
Hunting for the Right Low End for the Fugue Guitar.

Once you've found a frequency you like (I liked 257 Hz), pull the gain down to a reasonable level, and widen your *Q* out so it's not such a sharp peak of low end.

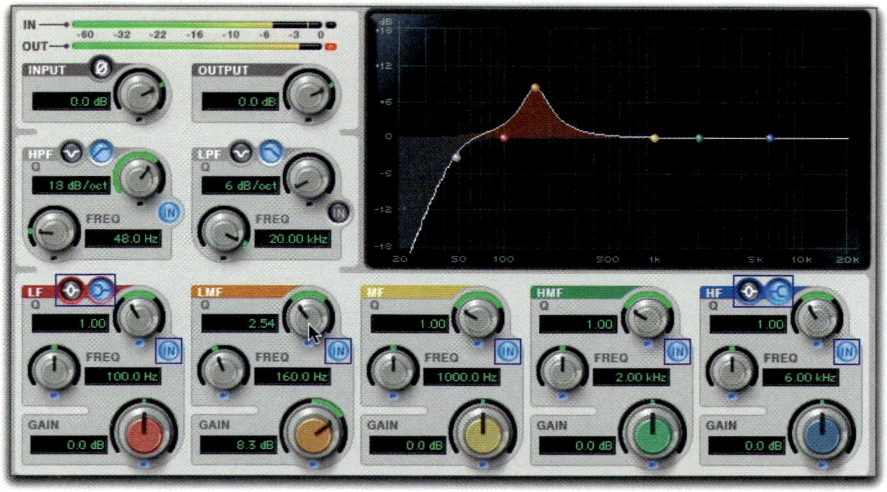

Figure 5.28 A Good Low-End Bump to Fatten Up This Particular Guitar.

Although that sounds nice, you may notice that by boosting this much low end, we're actually causing the output of our EQ to clip. Even though I don't hear an audible clip, let's turn down our Output knob until our output volume equals our input volume.

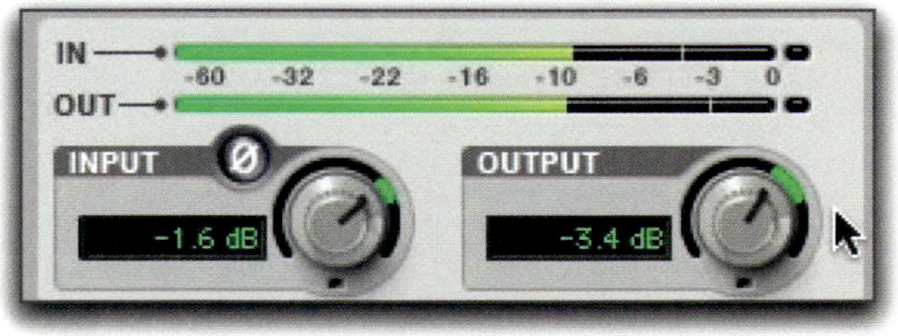

Figure 5.29 My EQ Ended Up at –1.6 In and –3.4 Out.

I actually end up pulling the output and input of the EQ just a bit to get safe levels.

Improving Stereo Image with EQ

This actually brings up a great point when talking about EQing two or more similar sounding tracks. In order to be able to make space for both instruments in a mix, we have used *complementary* EQ to boost a range in one, while cutting that same range in another. This has the effect of minimizing masking (the hiding of one instrument by another with a similar frequency range), and giving each guitar a bit of its own space in the mix. Try to always do this when mixing songs with multiple acoustics, electrics, keyboard tracks, and so on.

Another use for this complementary EQ trick is to make a mono track sound more stereo. For example, when mixing a track (recorded by an inferior engineer obviously) that has only one mono electric guitar, but sonically requires at least a stereo guitar, try this: Duplicate the audio track, pan them hard left and right, then boost 1k heavily on one track, while cutting it as extreme on the other. This will make them sound different enough to create a somewhat stereo effect. That's a freebie. Email me the next time that helps you in a session.

Testing the Stereo Guitar Mix

Alright, with all your guitars roughed in, EQ'd, leveled, and panned, jump back to bar 49 and give it a listen with the AIR Stereo Width plug-in bypassed, to listen to the width you've built into the track. We've only barely begun mixing, and things are already making dramatic improvements. Check the image below or download the session files to see any level tweaks we've made along the way.

Figure 5.30 The Mix Is Coming Along Nicely.

In This Chapter

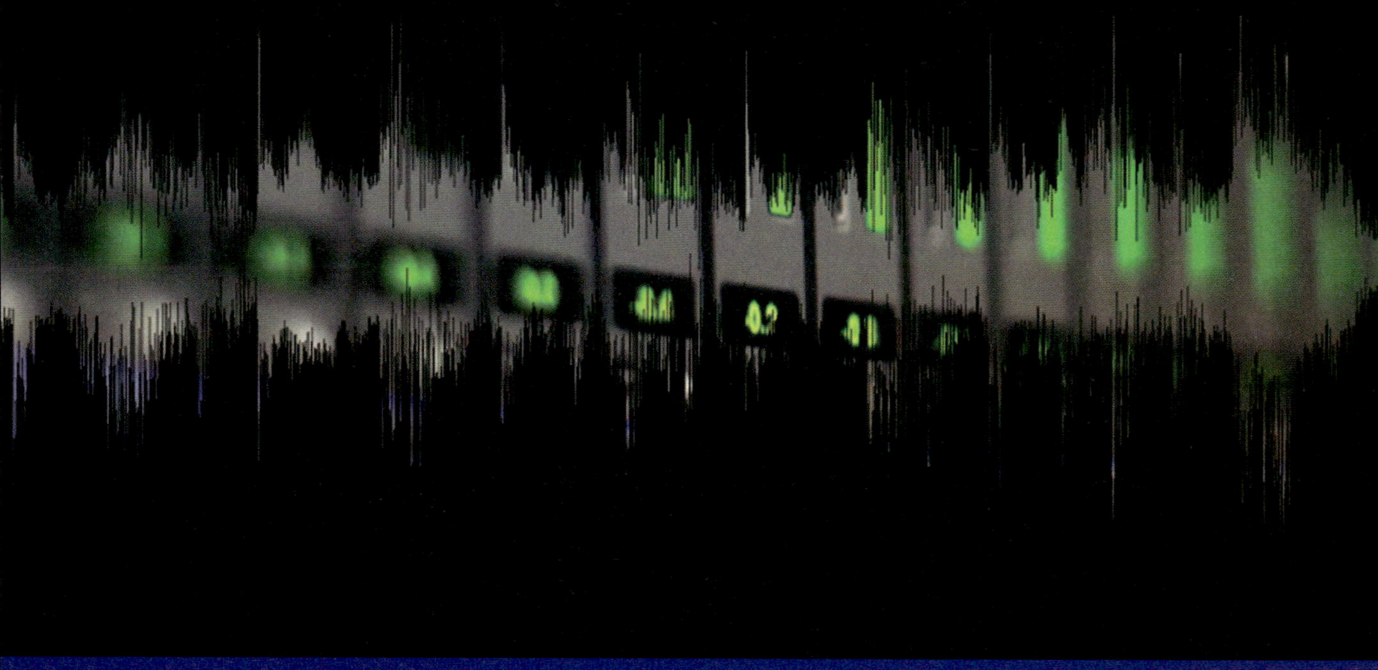

6

Bus Mixing

I know it feels like there's been a lot of work up to this point, but we're finally ready to start really mixing this track. As we've talked about previously, most great mixers develop workflows or templates that they're able to import into their session and reference to allow them to stay in the creative mode, rather than the "hunt for a good delay and reverb combo" mode. Of course, no matter how good the EQs and effects on your template are, they'll never work for *every* guitar, snare, or vocal sound you put them on. That said, rather than starting with one of my existing templates, I want to show you how to build your own to enable you to create faster, better mixes over time. Let's start by looking at the bussing structure we'll create to mix through.

Setting Up Submix Busses

Select the master fader at the far left side of your session and create four new Stereo Aux Input tracks (or use the shortcut). By selecting the master fader first, Pro Tools will know to put the new tracks immediately below or to the right of the selected track.

Figure 6.1 Creating Submix Aux Inputs.

Rename them to something memorable, like Vocal Buss, Band, Drums, and Gtrs, then let's get them set up for mixing.

© 2011 Nathan Adam. Published by Elsevier, Inc. All rights reserved.

(a)

(b)

Figure 6.2 (a, b) Setting Up Your Submixes.

To do this, we're going to select all of the drum tracks, route them out a stereo bus, then bring that bus into the Drums Aux Input. The fastest way is to click to the left of the Drums group in the groups window, then while holding our classic Shift+Alt shortcut, route any of the drums out of Bus 1 and 2.

Because Shift+Alt is the modifier that effects "All-Selected" tracks, all of your drums have been automatically set to go out Bus 1–2 together…and it only took two clicks instead of eight.

Now, because Bus 1–2 isn't terribly descriptive, let's rename it by right-clicking on the output selector and choosing Rename. In the dialog box that pops up, name it drum bus.

(a)

(b)

Figure 6.3 (a, b) Setting Up Our Drum Bus.

Now go to the input selector on your Drums Aux Input, and set it to the Drum Buss being fed by all of your drum tracks.

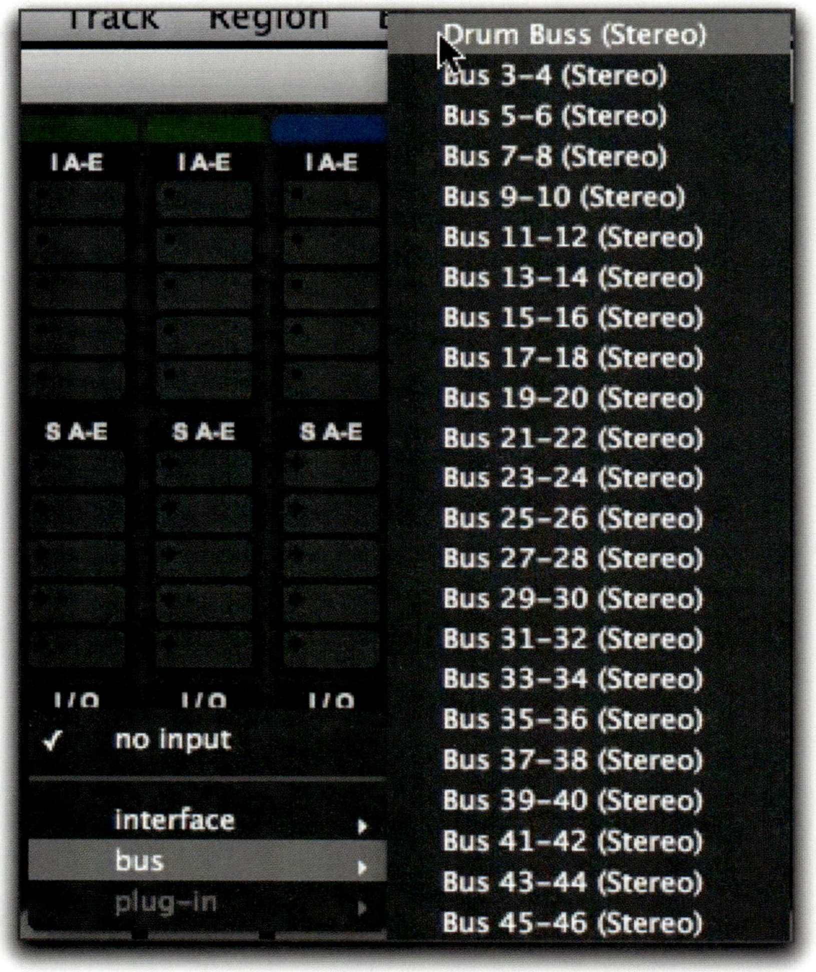

Figure 6.4 Bringing the Drum Bus into Its Submix Aux.

This has resulted in all of your drum tracks going *out* Drum Buss, which is coming *in* to a Drum Aux Input … effectively making the Drum Aux track what we call a "submix" of the drums.

Figure 6.5 All of the Drum Tracks Are Going Out Drum Buss, Which Is Coming into the Drums Aux.

Now that you understand the process, we're going to do the same thing with all of the guitar tracks.

1. Select all of the guitar tracks,
2. Use Shift+Alt and route any of the guitars out Bus 5–6,
3. Right click on the output and rename Bus 5–6 to Gtrs (or Guitar Buss, whatever floats your boat),
4. Assign the Gtrs Bus to the input of the Gtrs Aux track.

Finally, with all of your drums submixed to an Aux, and all of your guitars submixed to an Aux, let's submix both of those together by routing them out Bus 3–4, renaming it "Band Bus," and bringing it into the Band Aux track.

Figure 6.6 Woah. That's a Lot of Bussing.

For those new to mixing through busses, that may seem like a lot of work, but believe me, we're not even close to finished, and this is pretty much the minimum amount of bussing one will do in a mixing session. The only reason we don't have more submix Aux tracks is because this rock song only has primarily guitars and drums. If there were groups of keys, percussion, background vocals and sound effects, they would each be brought into their own submix and then sent onto the "Band" Master submix as well.

If you've done all of this correctly (use the image above as a guideline), you should now be able to mute the "Band" Aux track and hear nothing … because all of the instrumental tracks are ultimately going through this one, lone track. Or, if you mute the Drums Aux or the Guitars Aux, those should also go away.

Figure 6.7 The Band Track Is Now the Instrument Master Mix.

Now that you're practically a bussing expert, route the vocal and filter vocal audio tracks outputs to Bus 7–8, bring Bus 7–8 into the Vocal Aux, and rename Bus 7–8 to Vocal Bus.

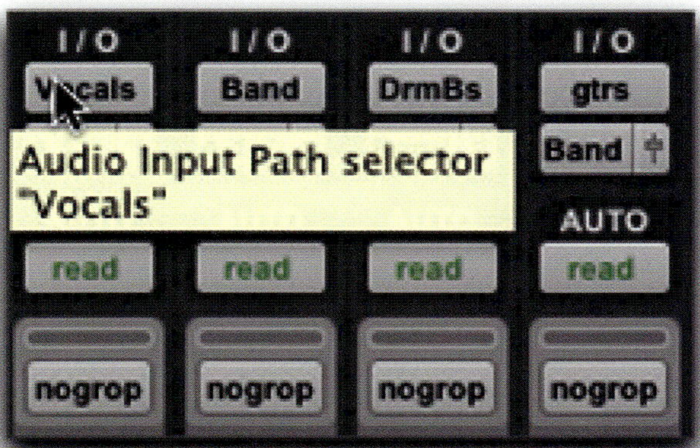

Figure 6.8 The Inputs of the Four Current Submix Masters.

Finally, before we can really use these busses to their fullest potential, try soloing the Drums Aux and listen to what happens.

Solo Safeing

You probably noticed that when you soloed the Drums or Guitars submix, you couldn't hear anything … because the act of soloing either of them muted all other tracks … including the Band submix that the Drums and Guitars submix is going out of. Hmmmm. What to do?

Figure 6.9 Soloing the Drums Submix Mutes the Band Submix … What Do We Do?

This is one of the perfect times for the Solo Safe function. Essentially, "solo safe" prevents a track from being muted when another track is soloed. We often use it on reverbs and effect returns so we don't have to constantly solo them every time we're trying to listen to our snare track in solo.

So just hold the Command (Win: Ctrl) key and click the solo button for each of the aux submixes and you will see them go gray. Additionally, they won't be muted whenever you solo another track.

Figure 6.10 The Benefits of Solo Safe.

Now that we've got our instrument submixes in place, we're going to create three more mono Aux inputs to be used as reverb and effects returns into our mix. Hit Shift+Cmd+N (Win: Shift+Ctrl+N) and create the three mono Aux Input tracks, then rename them to Dverb, Gverb, and Vverb (for drums, guitars, and vocals).

Figure 6.11 Our Collection of Aux Input Tracks Grows.

Sequential Inputs Trick

While they're all selected, let's assign their inputs to three sequential busses using a single keystroke. Since we know that Shift+Alt affects "All Selected" tracks, let's press them, but add the Command (Win: Ctrl) button to the mix (Mac: Shift+Alt+Cmd, Win: Shift+Alt+Ctrl) and select Bus 9 for the Dverb input. Finally, right-click and rename each of the busses to match their track names and solo safe them.

(a) (b) (c)

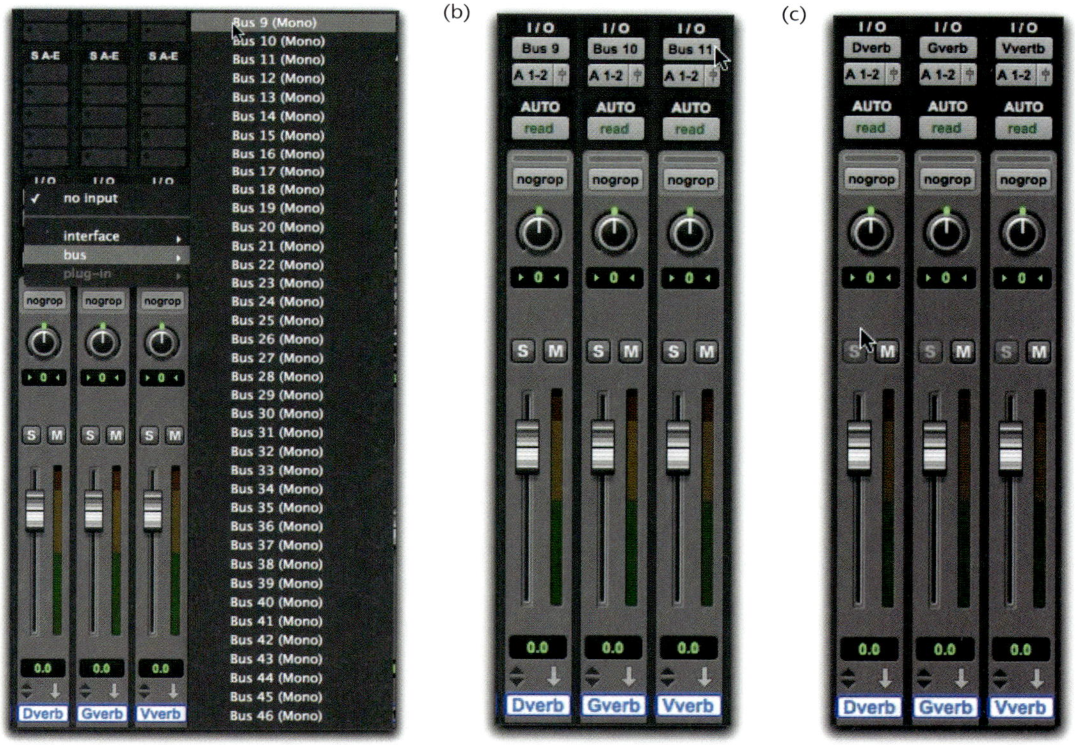

Figure 6.12 (a–c) The Shift+Alt+Cmd Modifier Sets All Selected Inputs Sequentially.

This is a great shortcut when you're trying to set the inputs or outputs for all of your tracks when coming from or going to a console, as it gets pretty tedious to individually route 48+ track inputs and outputs. Using Shift+Alt+Cmd, all the I/O can be set with two clicks.

Using the same technique, hold down Shift+Alt and add a Mono/Stereo Dverb plug-in to the three reverb aux tracks.

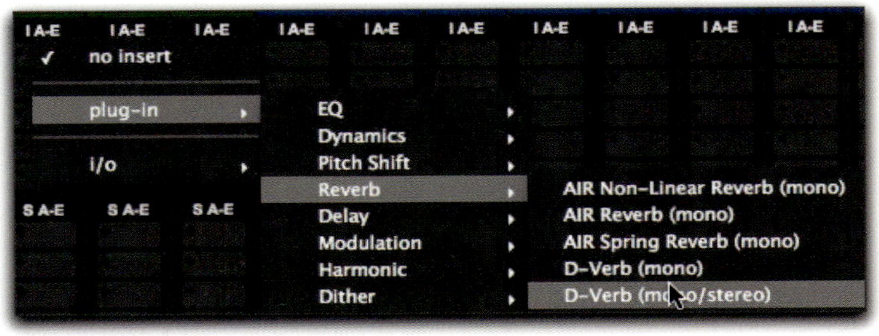

Figure 6.13 Quickly Adding the D-Verb Plug-in to Three Reverb Returns.

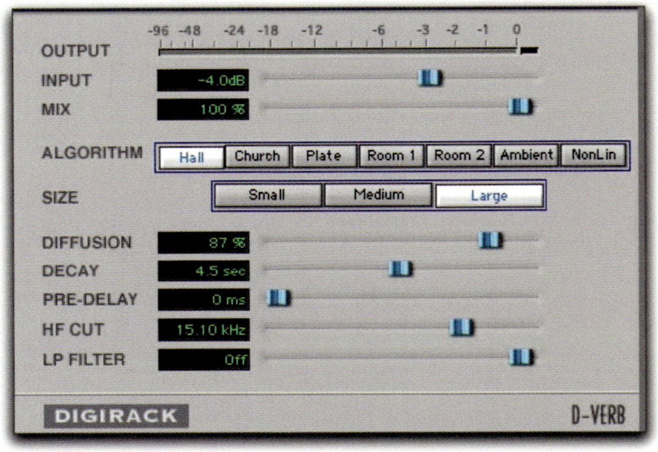

Figure 6.14 A Quick Reverb That We Can Tailor to Our Liking.

Reverb Returns

To finally get some reverb, we just need to route the appropriate tracks out a send to their chosen reverb bus. Let's start with sending the snare, Hi Tom, and Low Tom tracks. You can select all three and use a Shift+Alt+Click to instantly create sends to the Dverb bus on all three tracks.

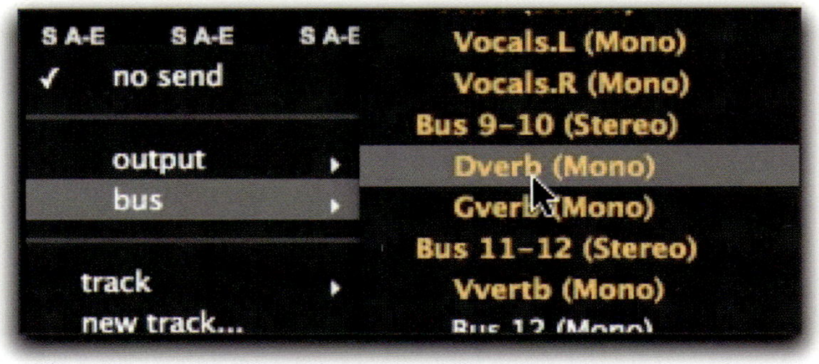

Figure 6.15 Let's Add Some Size to These Drums.

Depending on how your Pro Tools preferences have been set up, your sends may have started turned all the way down to negative infinity (wow, that's pretty quiet), resulting in no reverb. If that's the case, let's set a preference in Pro Tools that will speed up our workflow buy instantiating all auxiliary sends at unity level. Go to your Setup menu>Preferences>Mixing tab and uncheck Sends Default to "-INF." Now your sends will all start at unity level.

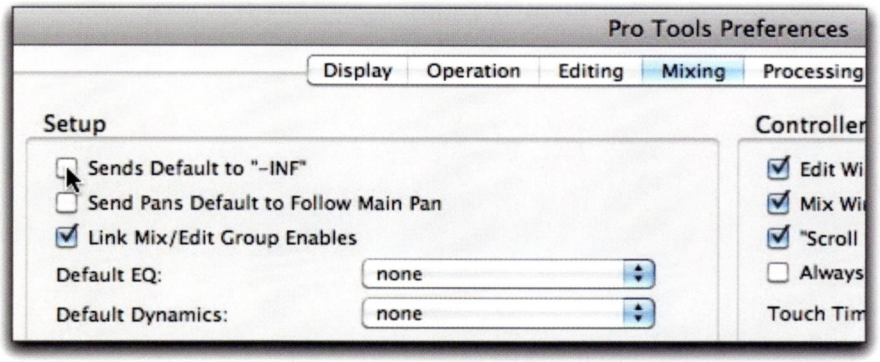

Figure 6.16 It's Good to Know the Preferences That Affect Your Daily Mixing Routine.

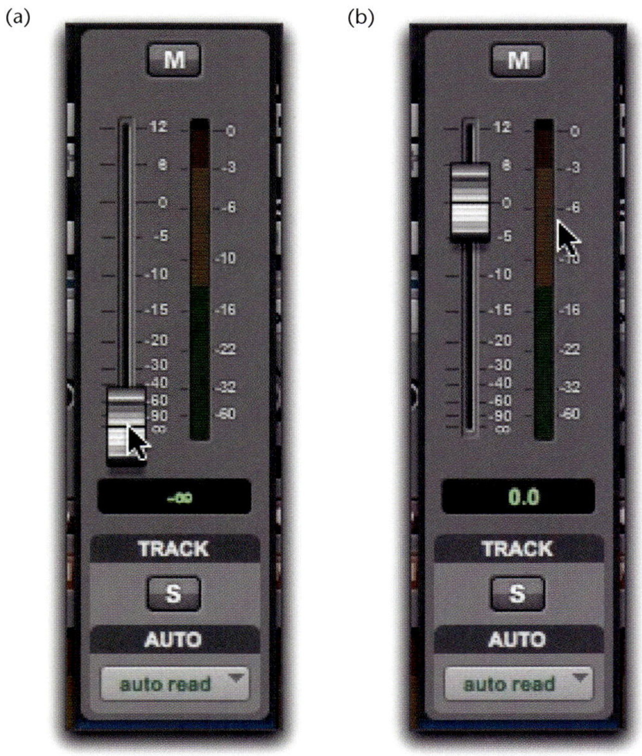

Figure 6.17 (a, b) Alternatively, If You Don't Mind the Extra Clicks, You Can Alt+Click Any Fader to Instantly Set It to Unity Gain.

Drum Verb

Now open up the Dverb plug-in on your Dverb track and let's find a sound that adds a bit of size and a modest tail to our drums. For our drum reverb, let's start with a plate style reverb. With Dverb open and the snare drum soloed, use the same plus and minus buttons to toggle through the various plate reverbs until you find one that you like. I've settled on Short Plate with the settings seen in Figure 6.18 below.

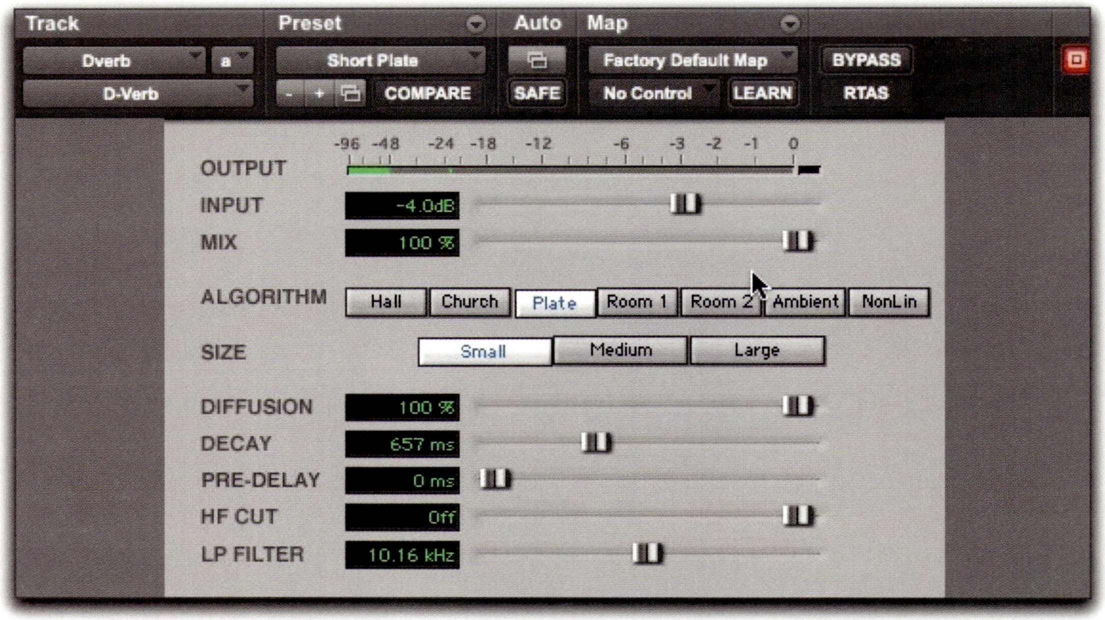

Figure 6.18 A Good Starting Snare Reverb Sound.

Setting Good Reverb Levels

Now let's start blending the reverb level of our drums with a simple trick that works in so many aspects of mixing.

1. To start tweaking our levels, un-bypass the AIR Stereo plug-in to make our mix mono again, which will help us hear phase and level problems as we mix. A quick way to do this is to simply Command+Click (Win: Ctrl+Click) the insert itself, which will toggle between a bypassed (dark blue) and un-bypassed state.

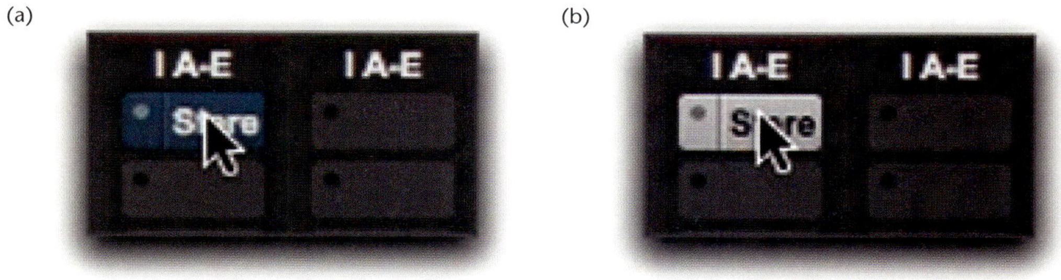

Figure 6.19 (a, b) Command+Click a Plug-in to Quickly Toggle It On and Off.

2. Pull the level of your Drum Verb Aux track all the way down, then slowly pull it up as the track is playing until you feel like you've got a sufficient amount of reverb audible in the mix.
3. Mute and unmute the Drum Verb track to see if you miss the sound of the reverb in the track when it's unmuted. If you don't miss it, it's probably not quite loud enough. If the change is too drastic, then it's probably overdone (unless that track calls for a giant reverb obviously).

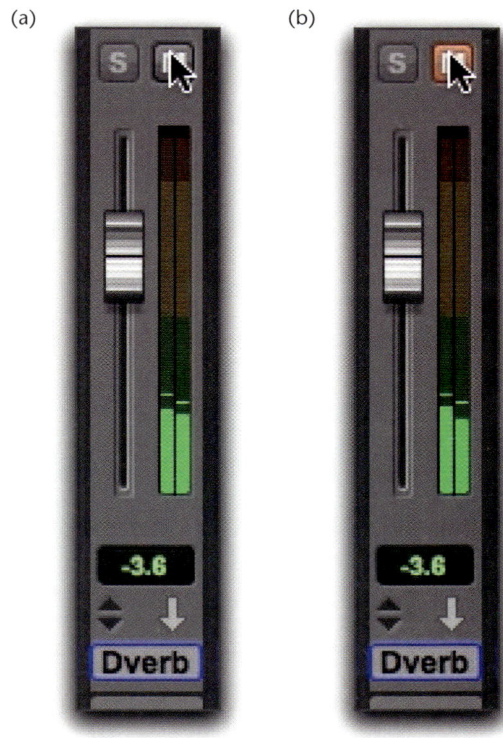

Figure 6.20 (a, b) Mute and Unmute Effects to See If You Miss Them in the Mix.

After a little tweaking, I like the Drum Verb return between –3 and –4 dB.

In This Chapter

7 Drum Mixing

Great! Now, it's time to start really mixing our Kick and snare tracks. One approach that is common in mixing is to find a solid pattern in the middle of the song (often on a chorus), where we can set up a quick loop of Kick and snare that can just play, repeatedly, while we EQ and compress them. I've found a couple bars around 23 and 24 that work well. So switch on Loop Playback mode under the Options menu (Mac: Shift+Cmd+L, Win: Shift+Ctrl+L) or just right-click the Play button on your Transport bar and choose Loop. The Play button should now have a looping arrow around it.

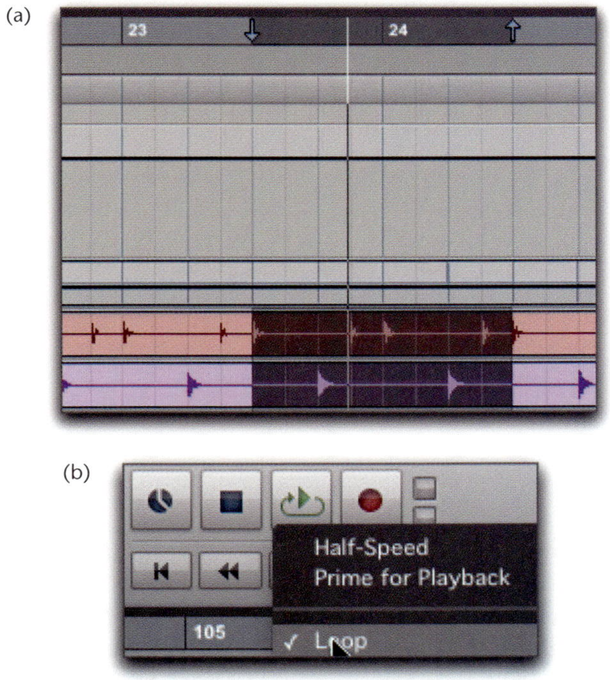

Figure 7.1
(a, b) Solo the Kick and Snare and Loop a Few Bars to Start EQing.

© 2011 Nathan Adam. Published by Elsevier, Inc. All rights reserved.

EQing Kick

Now that you have the system for EQing down, quickly drop a seven-band EQ on the kick drum and solo it in the mix. I feel like it needs a bit more attack, so let's use the same approach as we did on the fugue guitars to find the frequencies that add attack on this kick sound. I started with a wider EQ this time in the Hi-Mids (HMF) and swept around until I felt the attack on the kick start to cut through more. I ended up with about 7.6 dB at 1.94 kHz with a 0.65 *Q*.

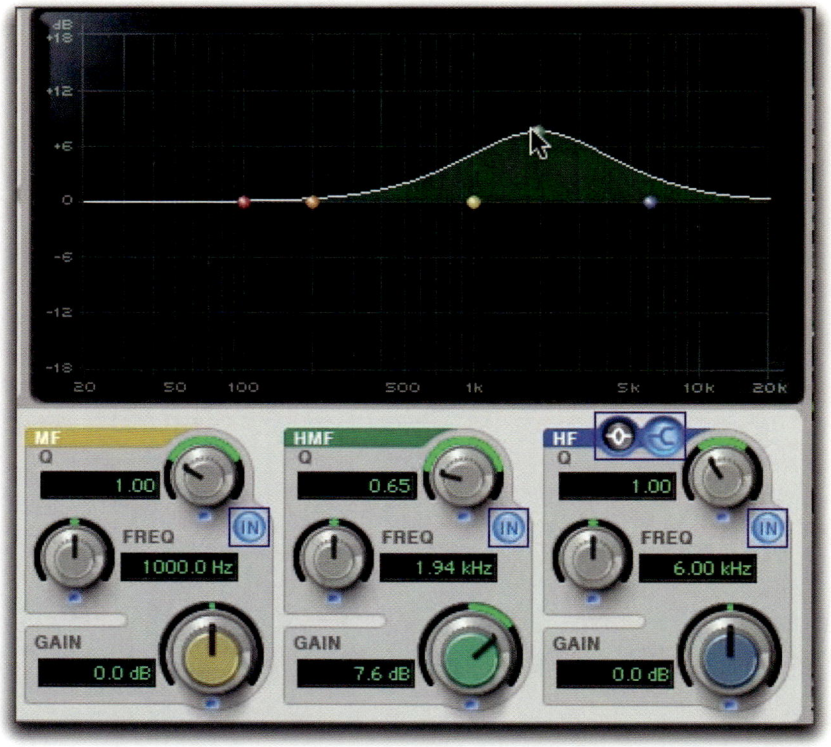

Figure 7.2
Adding Attack to a Kick.

Let's use the same approach as we did on the guitars to clean up the low end of the kick, and then add a bit of thump to it after we've removed the low junk. Start by switching in your HPF with a steep slope of 24 dB/octave and roll it up, until the kick starts to get thin. Then back it down, until you're not audibly hearing the lows go away. Mine ends up around 29 Hz.

Then, let's switch our low-frequency (LF) EQ in and toggle it from the Shelf style EQ to a bell shape. Now boost your gain with a high *Q* and sweep around the lows, until you find the sweet spot, where this kick drum get's just a bit of extra thump. Mine ended up around 41 Hz.

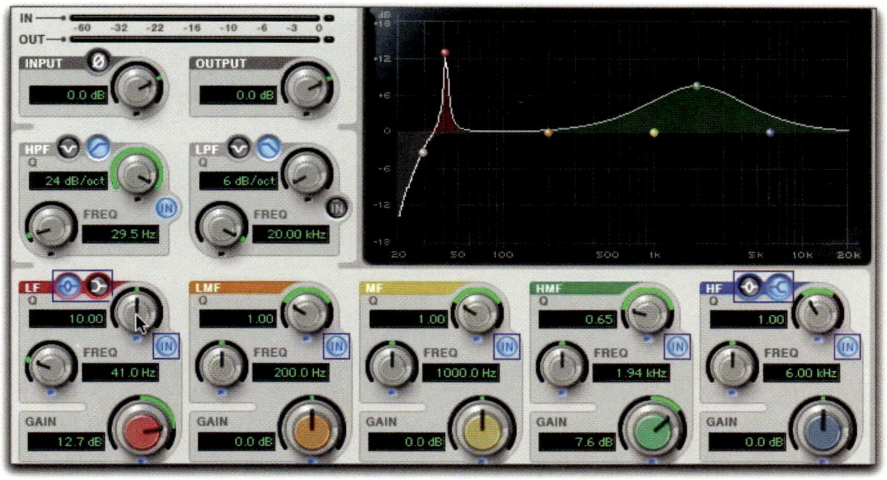

Figure 7.3
Finding the Low-End Sweet Spot on the Kick.

While we're at it, we almost never add large spikes of EQ with narrow *Q* as it tends to introduce weird phase sounds to the mix. So, after the best frequency for this kick is identified, widen the *Q* a bit and pull your gain down, until you're still hearing your nice bass boost, but it's not as extreme. I settled on a *Q* of 7.7 and 5.1 dB of gain.

Finally, I'm hearing a little boxiness in the kick that needs to be removed with a low–mid frequency (LMF) EQ. So grab the orange EQ dot and set a medium *Q* of around 3.85, and then pull it down between –6 and –12 dB, and sweep its frequency up and down while listening for the track to get punchier. It's often easier to hear the difference if you start with extreme gain changes, and once you've found the offending frequency, set the gain reduction at a more modest level.

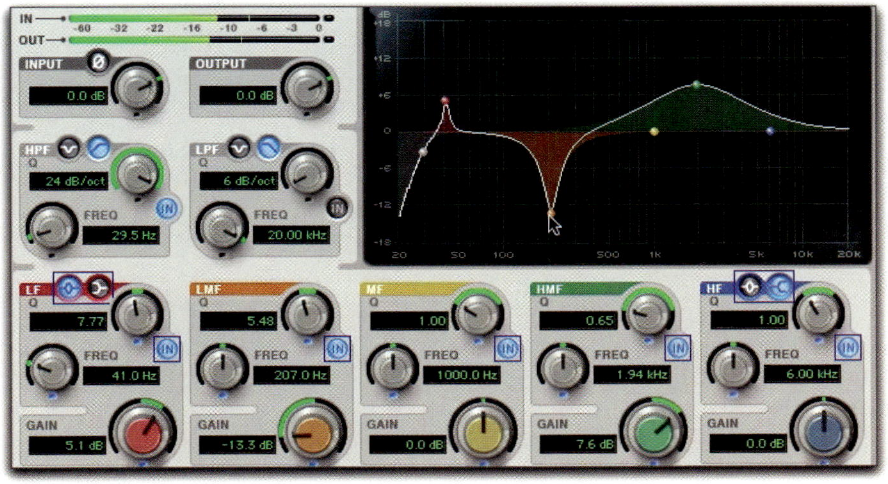

Figure 7.4
I Found Around 207 Hz to Be a Source of Boxy Sound on This Kick.

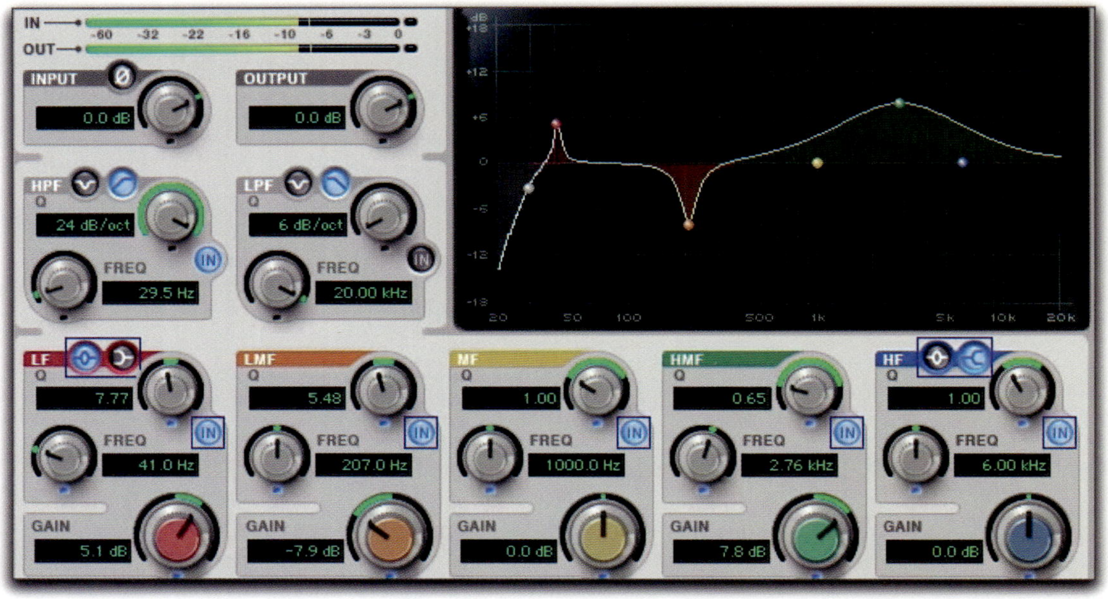

Figure 7.5 Once You've Found the Right Frequency, Bring the Gain Back Up to a More Modest Level.

Now, one of the hardest parts to get used to when mixing is just how short of a memory your ear has. So it's easy when EQing to forget what a dramatic change you're making to the instrument after you've been listening to it soloed for a while. That's why, especially when you're learning, you should regularly reach up to that BYPASS button and toggle it in and out, several times, while the track is playing. Do it now with the kick drum, and you'll hear just what a dramatic difference we've made to the track.

(a)

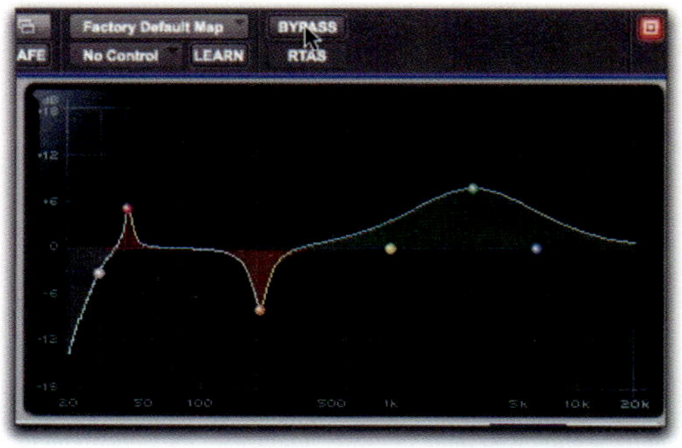

(b)

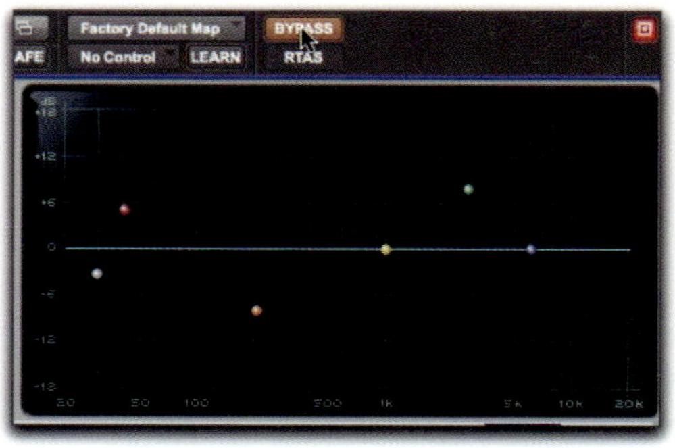

Figure 7.6 (a, b) Bypassing the Kick EQ for Comparison.

Drum Compression

For a kick drum that was cut by the band in their garage, we've taken the first step toward getting our preferred drum sound, by using EQ to "fix" or compensate for the deficiencies in the recorded sound. Now that we've fixed the general sound, let's move on to the enhancing stage by adding a bit of compression. Go ahead and put the bombfactory BF76 plug-in (Plugins>Dynamics>bombfactory) after the EQ on the Kick track.

Let's start with the same technique that we used way back on the acoustic guitar by setting our slow ATTACK and our fast RELEASE to see if it tightens up the low end on the kick. As always, switch the BYPASS button in and out to compare the sound of the effected and uneffected track.

Figure 7.7 A Quick and Gentle Compression on the Kick.

I'm liking the way it tightens up the low end without sounding too aggressive; so in order to stay in the fast-creative mode, let's move ahead making another wave of quick, cumulative changes to our mix.

EQing Snare

Let's move over to the snare, and to be efficient, copy (Alt+drag) the EQ and BF76 over to it from the Kick track. Now open the EQ and quickly reset the EQ curves by Alt+clicking the gain knobs along the bottom.

(a)

(b)

Figure 7.8 (a, b) Alt+Clicking Resets Any Parameter to Zero.

First things first. As we've done with most tracks, let's filter out the junk from the ultralow end of the snare. As you've done it so many times at this point, you know the process. My EQ ended up at 124 Hz with 18 dB/octave slope.

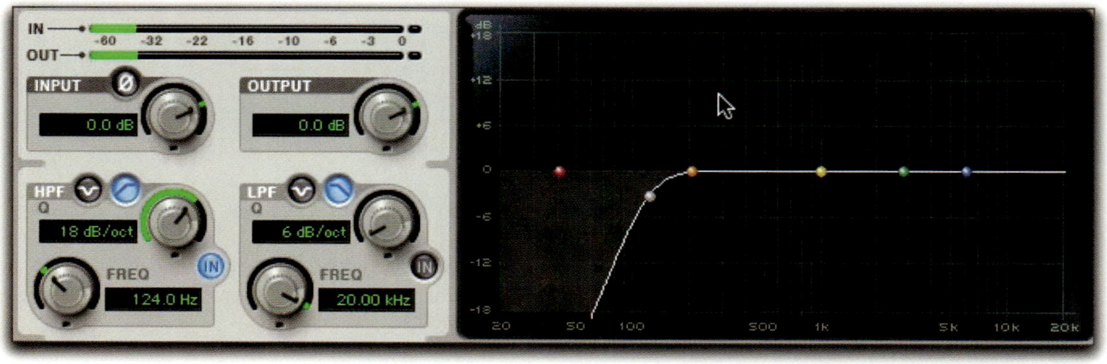

Figure 7.9 Filtering the Snare.

Second, every snare drum has a fundamental low-end frequency that we want to find and accent. Grab your LMF EQ and boost it with a tight *Q* of ~5.5. Start sweeping the frequency around listening for the snare to start punching you in the gut a bit harder. If you hit around 237 Hz, you'll hear it sounding a little too "honky," but down at 187 Hz, there's a low-end punch that just sits really well with this snare drum. Pull your gain back down to a more reasonable 5–6 dB, and then let's add some teeth to this snare drum.

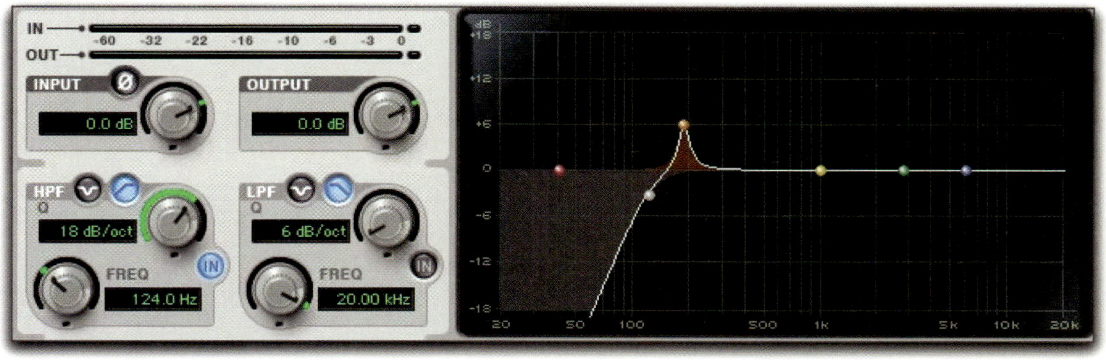

Figure 7.10 Adding Some Fundamental Lows to the Snare.

For most snares, 5–6 kHz is where they really start to "sing," so I started with a wide boost (*Q* of 0.65) in that range with the HMF EQ and then swept the frequency and gained around, until I found the sizzle for this snare. I ended up at exactly 5 kHz with an 8-dB boost. While we're at it, I'm feeling like even a little more of that low-end thump could be added, so let's turn our LMF gain up to 9.4 dB.

Oops! With all this additive EQ, it's pushing our EQ output level into the clipping zone. Let's back down the output gain to –5.1 dB, so that it's even with our input level. That's better.

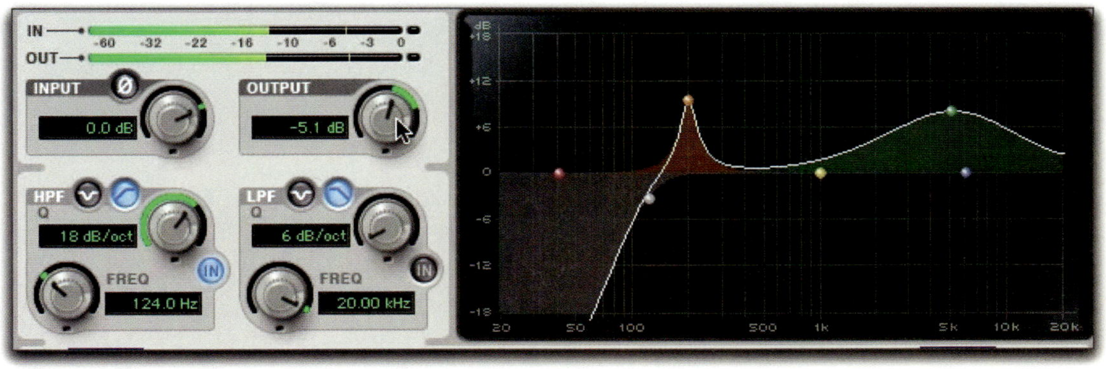

Figure 7.11 Setting the Output Gain to Match the Input Gain.

Here's one of the challenges of boosting so much of the low fundamental frequency on this snare ... it tends to get overemphasized in one of the harmonics of that frequency, making the snare sound a little muddy. To find the mud, use your mid-frequency (MF) EQ to make a tight boost and sweep around the harmonics of the 200-Hz frequency that we boosted with our LMF (400 Hz, 800 Hz, and so on). Sure enough, at 400 Hz, we've got a muddy sound that's unbecoming of our snare track.

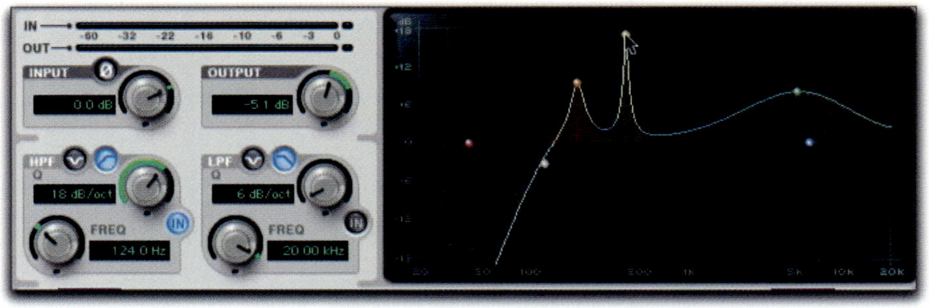

Figure 7.12 Identifying the Muddy Frequency.

As we've always done, now you want to pull that frequency gain down, until we're making a cut at the offending frequency. At around –2 dB, it feels like we've cut enough of the 400 Hz. Be sure to double-check it with the bypass/unbypass trick.

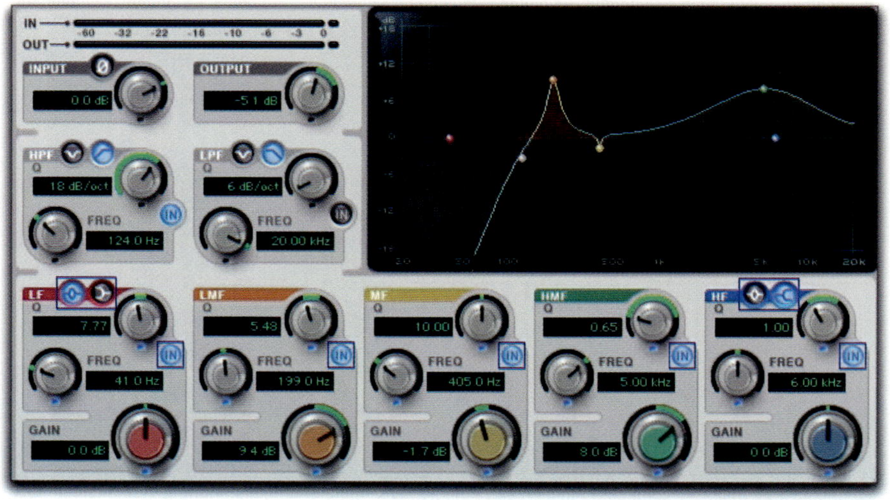

Figure 7.13 Our Final Snare EQ.

Excellent! Now that we're making progress through our drums, the one track we haven't attempted to bring into our mix is the room microphone over on the end. Temporarily turn off your drum group in the Mix window by pressing the "a" key, then try bringing up the room microphone to see if it's useful for the mix you're trying to accomplish. After listening to it, I can say that having worked with a lot of good room sounds in the past, I'm pretty certain that the sound captured by this microphone in a garage isn't one of them. This just goes to show you that you can't always use every track in a mix. So lest we accidentally add it in later, simply right-click on the room microphone and choose "Hide and Make Inactive."

Figure 7.14 Deactivating the Room Microphone.

Tom Processing

With that out of the way, let's mix on the Toms. Select a loop around the Low Tom at bar 20 and add a BF76 compressor to it. Using the same settings as before (slow Attack at 1, fast Release at 7), adjust the input level until you feel the compressor adding a little more punch to the Tom hit.

Figure 7.15 Somewhere around Input of 28, the Tom Really Starts to Pop.

The problem with adding this compressor to the Tom, however, is that when it brings the level of the Tom track up, it also brings up the background noise. One solution would be to apply a gate or expander to the track to try to minimize the bleed of the other drums into the Tom microphone. However, because he didn't play them very much in this track, it's going to be faster to just go in and edit or mute everything but the Tom hits from the track.

Figure 7.16 The Bleed in the Tom Track Needs to Be Removed.

Make sure that Tab to Transients is still on, place your selection just before the Tom hit at bar 20, and press the Tab key until your timeline selector is at the front of the Tom hit. Use a shortcut to separate the region, then select all of the audio before it and mute it (Edit menu>Mute Region) with Cmd+M (Win: Ctrl+M).

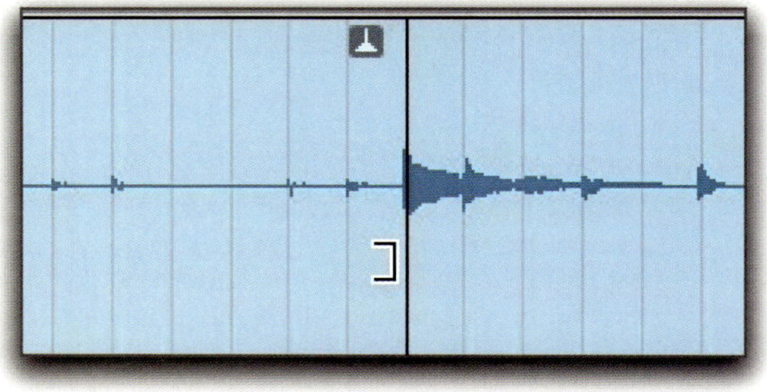

Figure 7.17 Tab to the Tom Hit and Separate the Audio Region Before It.

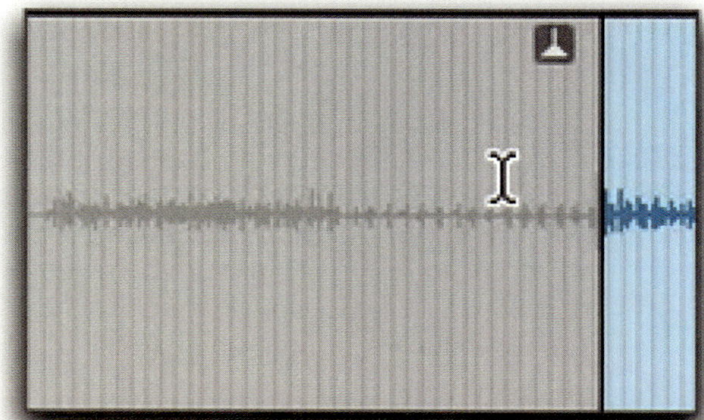

Figure 7.18 Muting the Bleed Audio Before the Tom Hit.

Now Tab to the end of the sustain for this Tom hit at 21|2, separate, and jump down the track to the next Tom hit at 77|3. Separate the region and mute all the audio between the two Tom hits.

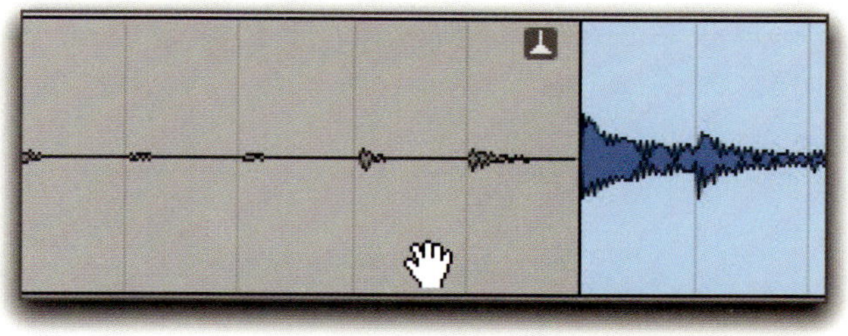

Figure 7.19 Muting the Audio between 21|1 and 77|3.

Go ahead and clean up the rest of this Tom track by separating and muting the audio between bars 87 and 109, bars 111 and 125, and bar 128 to the end.

Figure 7.20 Continuing to Clean the Tom Track.

After you've cleaned up the Low Tom track, spend a few minutes to do the same to the Hi Tom track, so that you'll be able to focus on mixing just the sound of the Toms, and not the sound of the bleed behind them. If you get stuck at any point, be sure to download the Pro Tools sessions to see exactly where I chopped and muted as a reference.

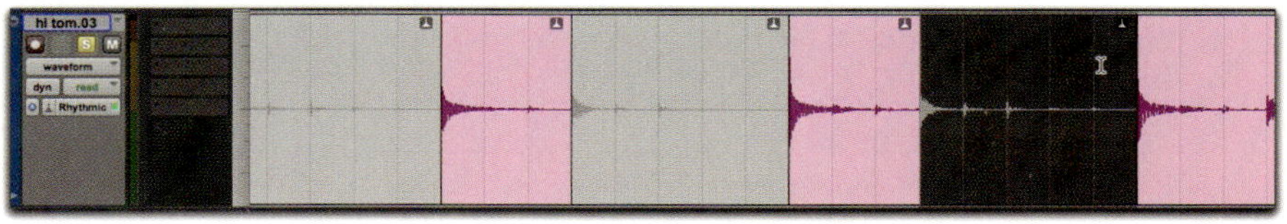

Figure 7.21 Cleaning the Hi Tom Track at Bar 80.

Never forget, it's not just one big trick that results in better mixes. It's 1000 small decisions that accumulate into a truly great track.

Now that they're clean, copy the BF76 from the Low Toms up to the Hi Tom track and jump to bar 125. You should have left a Tom fill section in here, and I'm hearing a kind of splatty, flammy Tom right at bar 126|1. No worries! Just copy the Tom beat from bar 125|1, and paste it right on 126|1. Just in case there's any clicks and pops to either side, go to Edit menu>Fades>Create, or hit Cmd+F (Win: Ctrl+F).

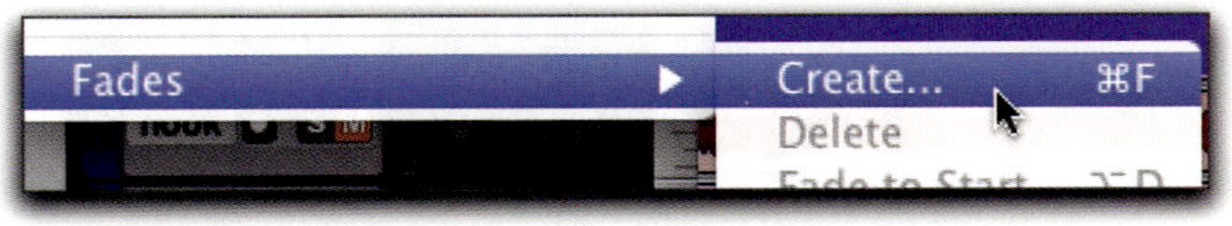

Figure 7.22 Creating Fades.

This will create fades on both ends of the selected region, eliminating any glitches in the audio.

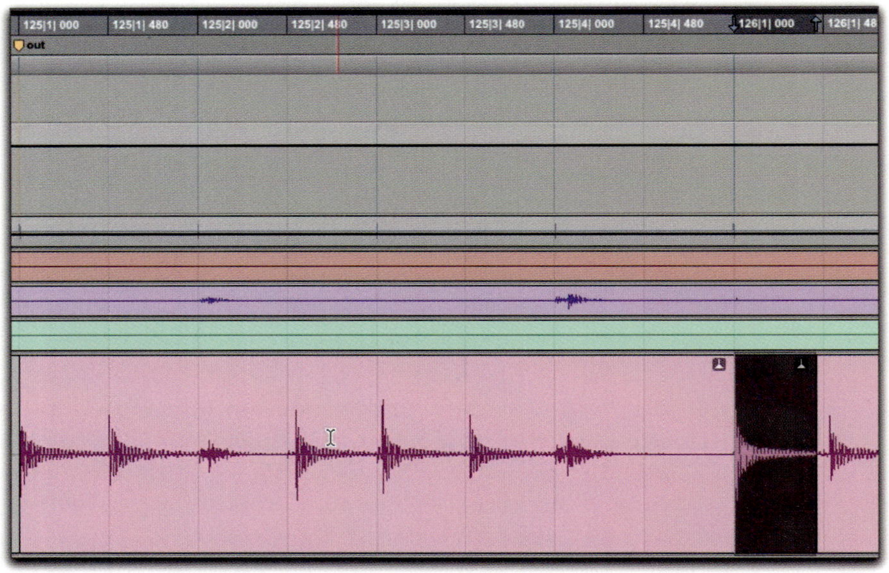

Figure 7.23 Hiding a Weak Tom Hit.

Alternatively, we can go through and use our Smart tool to add simple fade outs to all of the Tom hits.

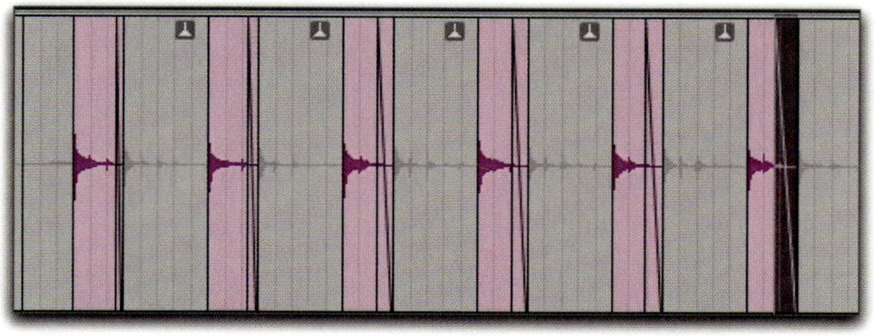

Figure 7.24 Fading Tom Hits after Cleanup.

Now as our Toms are mono with minimal bleed, take a minute to blend them in the mix by playing over the Tom section at bar 77 and adjusting their faders until they add to the feel of the mix, without the feel of overpowering. Mine ended up around –5.5 dB.

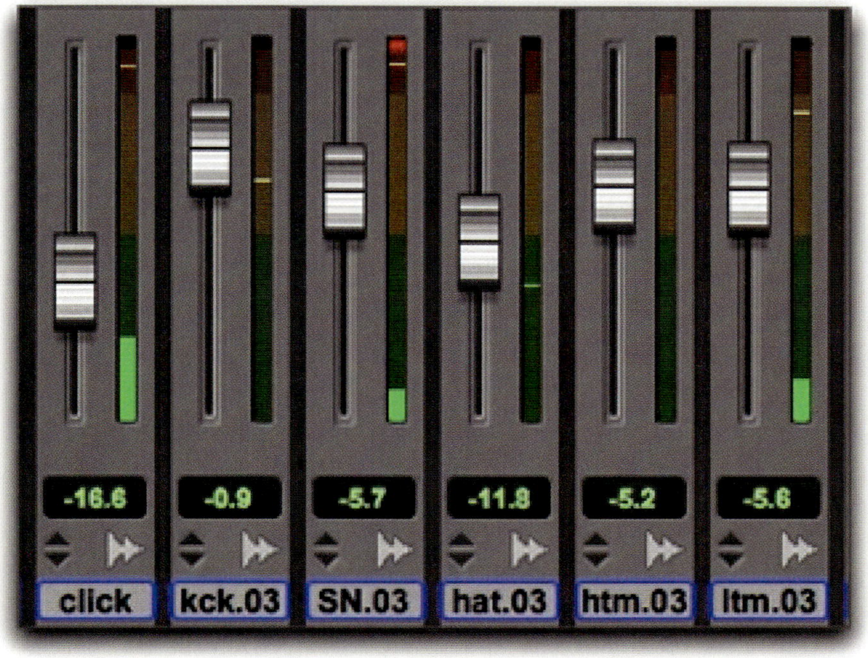

Figure 7.25 Our Levels so Far.

We're at a point in our mix where we get to start performing the smaller and smaller steps that make a good drum mix. That's right, we're at the point where we need to check the phase on the drums (for the nitpickers out there, I'm aware that we're actually checking polarity … but phase is the more commonly accepted expression, and what we'll use).

To start, move your overhead track up beneath your kick drum track, solo them both, and zoom in closely to visually examine the waveforms. Let's see how the phase of the Kick track is lining up with the overheads.

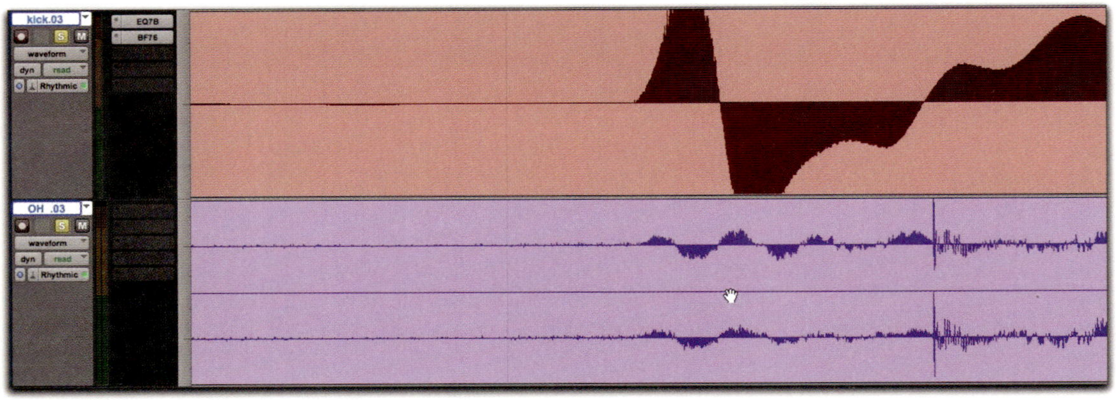

Figure 7.26 The Phase of the Kick and Overhead Tracks.

Phase-Checking Drums

One way we can find out about phase problems is when an instrument or track that sounds great in stereo disappears when being played back in mono over a radio or other mono playback system. Again, this is one of the reasons we are doing our early mixing in mono, so that we can hear and correct such problems along the way.

If you find that your tracks are out of phase, during the recording process, the simple solution is to go to Audiosuite menu>Other>Invert and invert the out-of-phase track, permanently, so that whoever mixes your track won't, accidentally, leave it out-of-phase (and thus disappearing from the mix during its radio debut).

(a) (b)

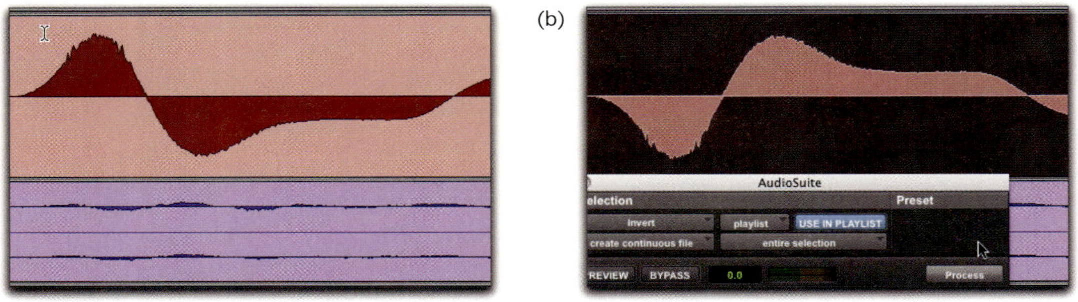

Figure 7.27 (a, b) Before and After Inverting the Phase.

Alternatively, if you're mixing the track and it has an EQ plug-in already on it, you can just use the Phase Invert button built right into it.

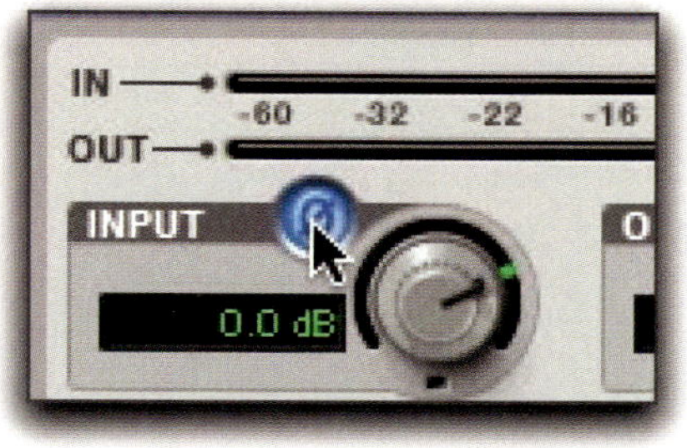

Figure 7.28 The Phase Invert Button When Turned On.

So to check our drums, let's actually solo the Kick and overhead tracks, and toggle the phase of the kick drum to see whether it sounds *best* as it currently is, or when the phase is inverted. Just because it appears to be in relative phase, doesn't mean it's the best. Trust your ears.

Automatic Delay Compensation

Earlier versions of Pro Tools LE lacked what is known as "Automatic Delay Compensation." As a result, every plug-in placed on a track actually introduced a tiny amount of delay … shifting that track just a little bit later in playback than the other tracks. While it was rarely enough to move a track audibly out of sync, it could often result in pushing tracks in a multimiked setup (like drums) out of phase … resulting in less clarity and definition in the high end.

Fortunately, Pro Tools 9 brought the feature from the Pro Tools HD systems down to the masses, and we no longer have to deal with inferior phase relationships caused by plug-in delay. Thanks Avid!

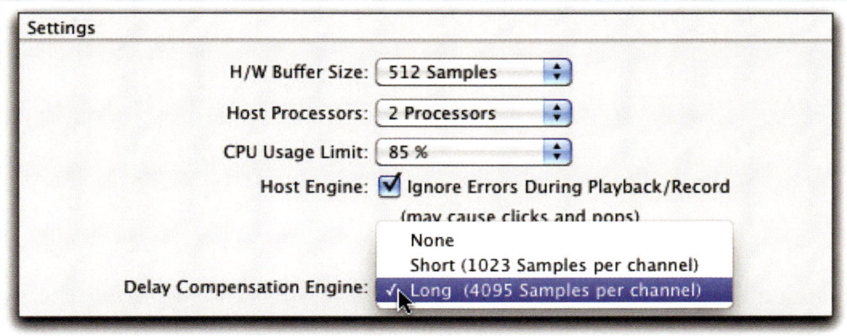

Figure 7.29 The Delay Compensation Engine Can Be Turned On or Off in Your Setup Menu>Playback Engine.

Loop bar 8 with the kick and overheads soloed, and toggle the phase switch on the EQ of the kick drum. What you'll hear is the low end without the Phase button pressed is a little bit thinner and tighter, and when the phase is reversed, the low end is a bit bigger and looser.

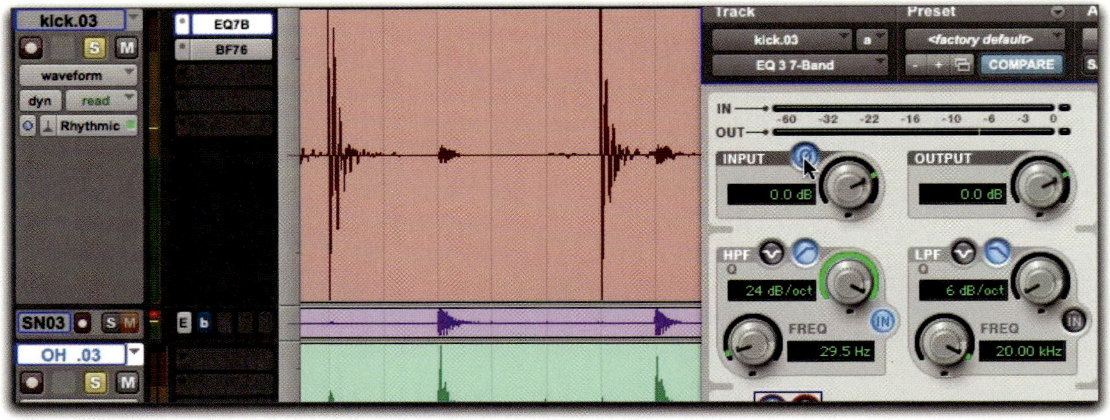

Figure 7.30 Toggling the Phase Reverse on the Kick Track.

I actually prefer the phase-reversed kick drum as it provides a better low-end punch on the track. Now, let's solo the snare against the overheads and do the same thing.

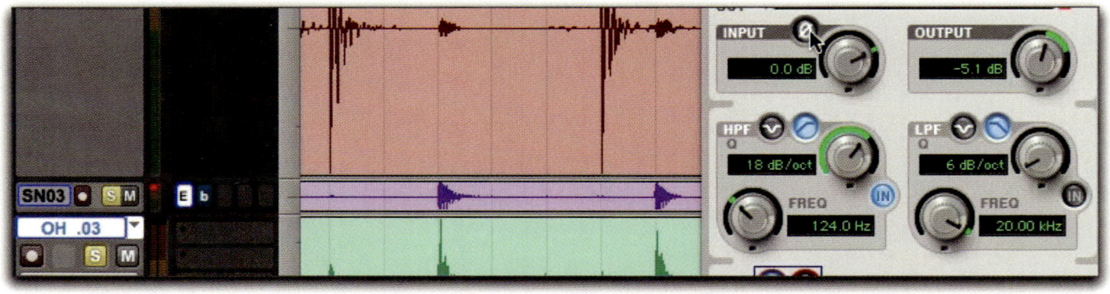

Figure 7.31 Phase-Checking the Snare Drum.

In this case, the out-of-phase snare sounds thinner and whispy, where the untouched snare sounds thicker. So for the snare, let's leave the phase reverse off (gray).

Let's do the same to the Tom tracks. Because they don't have an EQ, let's just add a 1 Band EQ3 plug-in and toggle the phase as we loop the Toms.

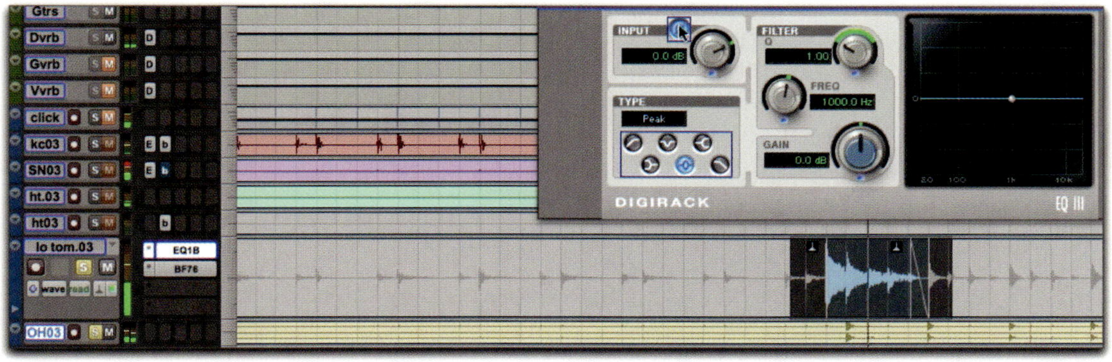

Figure 7.32 Checking Phase on the Toms.

I prefer the phase reversed on the Low Tom and without phase correction on the Hi Tom.

Moving ahead, because our overheads already sound pretty good, and our drums are now in their best phase, we're ready to try a few more drum-mixing tricks to make them really pop.

Capturing Safety Drums

The first trick we're going to show you is commonly called drum replacement. In fact, if you look way out at the end of the session around bar 205, you may have noticed recordings of a bunch of individual drum hits.

When consulting with a band, I generally tell them to always record a variety of individual hits of their drum kit, at multiple velocities, for possible sound replacement later.

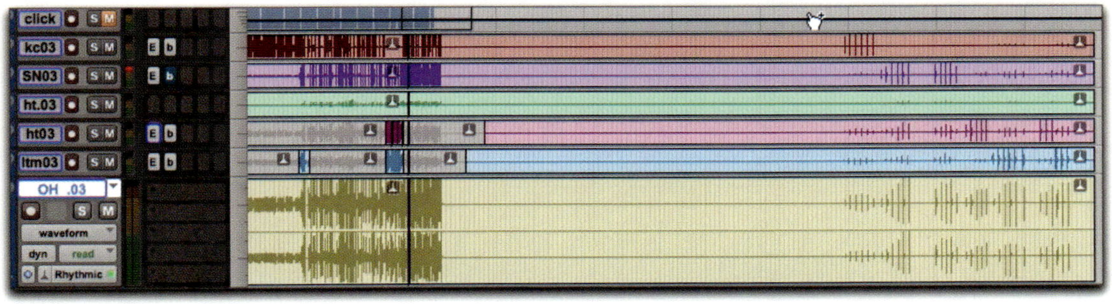

Figure 7.33 Drum Safety Hits Recorded at the End of the Session.

Especially for "real" sounding rock records, as opposed to say an electro-pop record, blending or replacing any missed hits with the actual, original drums tends to yield a more realistic sound. On a pop record that is looking for giant, almost comically processed drum sounds, I'll blend the drums with processed samples in third-party programs like Sound Replacer or Drumagog, but for raw, edgy rock like this, I would like to have the option of clear extra hits to replace with.

Manual Drum Replacement

Although I strongly recommend picking up either of those programs, it is definitely possible (if time-consuming) to manually replace all of the snare (or kick or Tom) hits with only Pro Tools 9 using our old standby, Tab to Transients. Let's take a quick look at the process.

Let's start by listening through the clean snare hits and choosing our favorite.

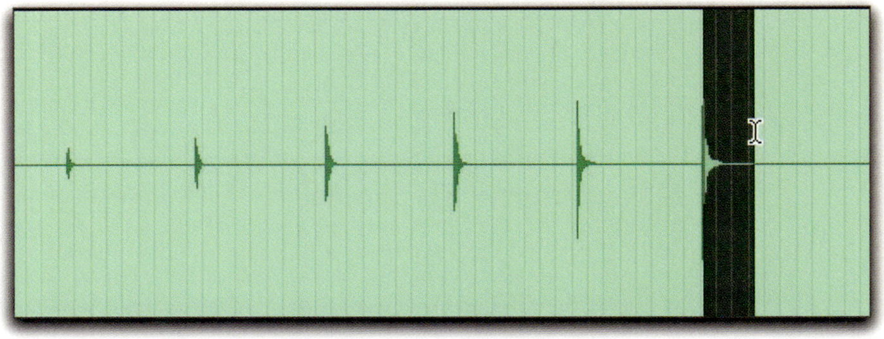

Figure 7.34 I Like the Last One for Its Punchy Sound.

Use Tab to Transients to move your selector right up against the snare hit, then Shift+drag to select from the front of the snare sound, until the sustain of the snare is past. Zoom in on the waveform, if necessary, to make sure you have the entire snare sound and aren't cutting off the attack or sustain.

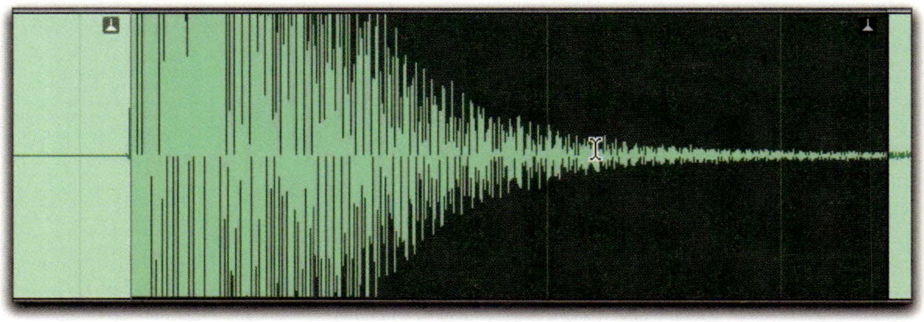

Figure 7.35 Grabbing the Entire Snare Sound.

Finally, add a natural-sounding fade to the end of the clean snare hit, select it, and consolidate it into a new audio file with Edit menu>Consolidate Region (Shift+Alt+3). This will create an entirely new audio file that can be copied and pasted throughout the session. Select the newly created snare sample and copy it.

(a) (b)

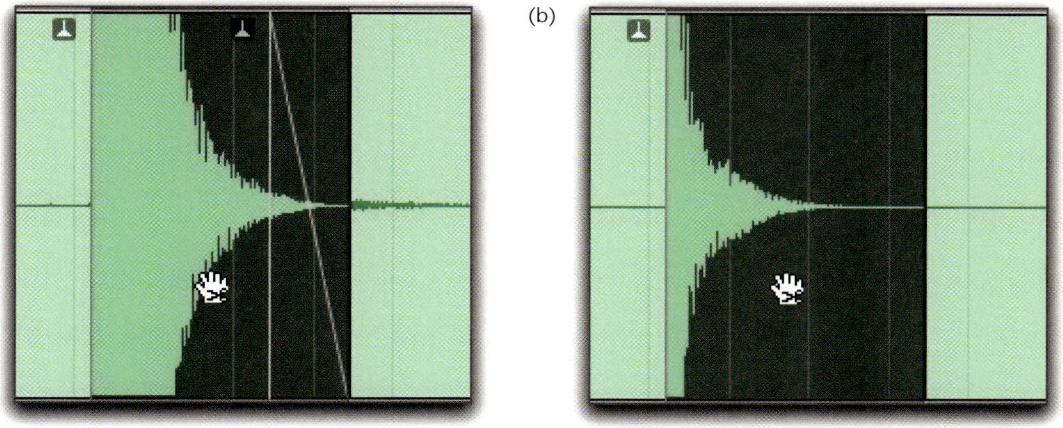

Figure 7.36 (a, b) The Final Replacement Snare Sound Is Ready for Copying and Pasting.

Now, just to be safe, let's duplicate a new playlist on the snare track that we'll be able to paste our newly captured snare drum to. Call it something like "Snare.trigger."

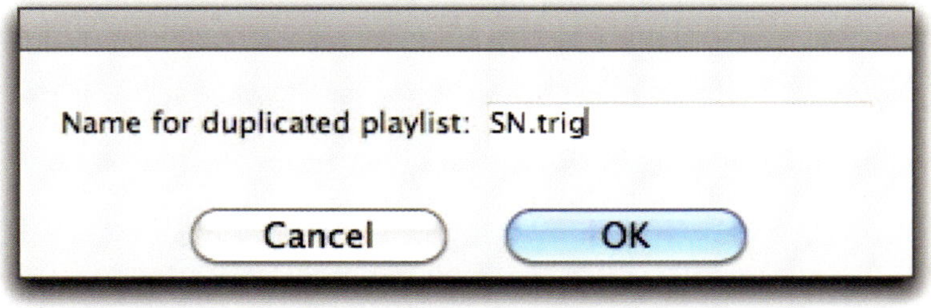

Figure 7.37 Making a Triggered Snare Playlist.

This will enable us to get back to our original drum sound if we manage to mess things up. Next, select the existing snare region and mute the entire thing (Mac: Cmd+M, Win: Ctrl+M). Even with the region muted, our Tab to Transients function will still work; so rewind to the top of the song and tab to the beginning of the first snare hit.

Well wouldn't you know it, our very first snare hit is going to be an exception to the rule. If you look closely, you'll see that this snare hit at 4|4 has a bit of a double transient … in other words, it's a flam.

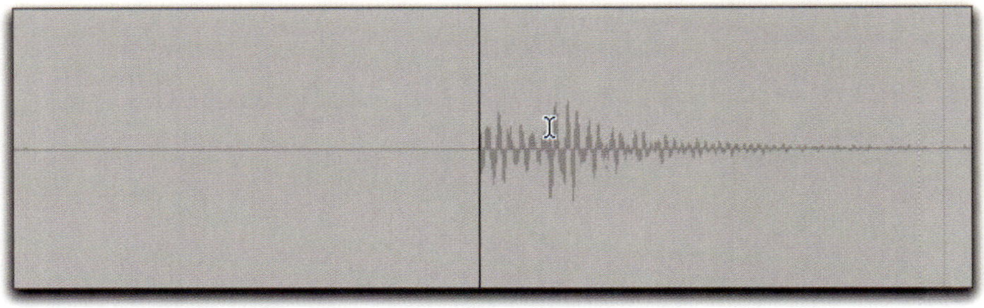

Figure 7.38
Looking for Flam Hits.

This is the sort of sound you'll need to look out for, when manually replacing a drum sound, as pasting a single-snare hit on the top of a flam won't sound right when combined with the flam sound present in the overheads (not to mention any bleed into the other drum tracks).

So, hit Tab again and Pro Tools should advance the selector to the next, larger snare transient, ready for you to paste in your earlier-copied clean snare. Once it's in, select the flammy snare in front of it and unmute (Mac: Cmd+M, Win: Ctrl+M).

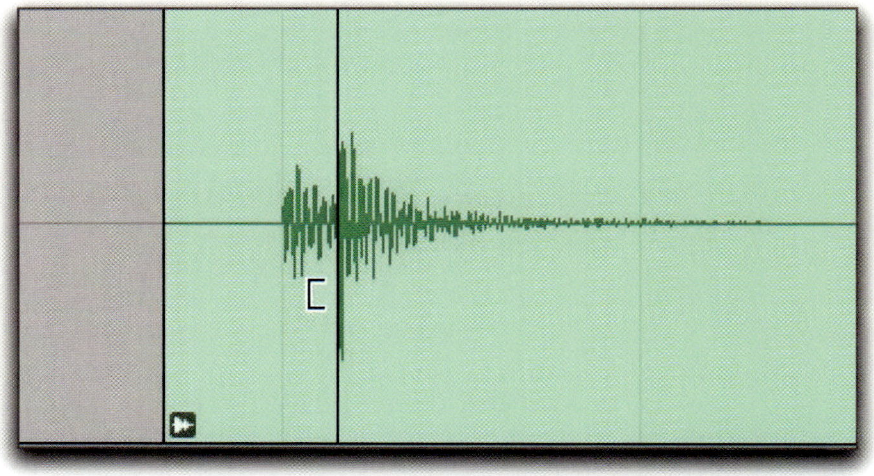

Figure 7.39
Dealing with Occasional Flams and Ghost Notes Is One of the Challenges of Manual Drum Replacement.

Now, you're ready to dive deep into the monotony of repetitive keystrokes as you Tab–Tab to the next snare transient, and Edit menu>Paste (use your shortcut!) the sampled snare drum over the top.

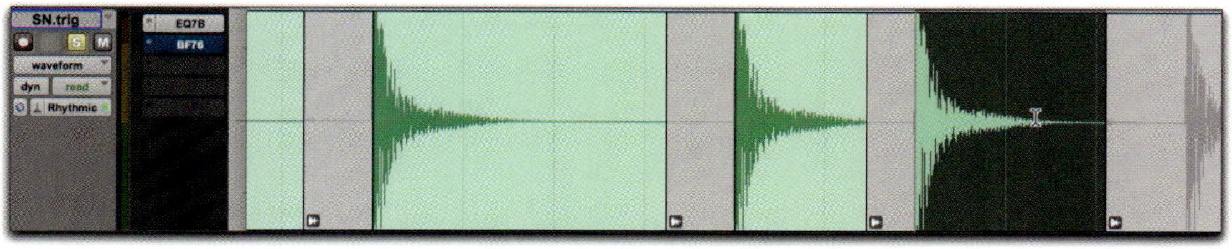

Figure 7.40 Manually Pasting Sample Snare Hits on the Original, Muted Snare Track.

When doing this, you'll want to keep a close eye on every new snare drum to make sure you're actually pasting single-snare hits over single-snare hits. Although manual drum replacement is relatively easy on a straightforward rock song (a full snare track can usually be replaced in 10–15 min), if the track contains a lot of ghost notes, flams, dynamic builds, or other snare trickery, an actual drum replacement program will be an order of magnitude more efficient at completing that task.

Parallel Drum Compression

As a result, the technique we're going to use for this session is known as parallel compression, or also New York style compression. To set it up, select your Drums aux submix track and duplicate it (Track menu>Duplicate or Shift+Alt+D). Then rename it PDrums or something similar to help you identify it as the parallel compressed drums.

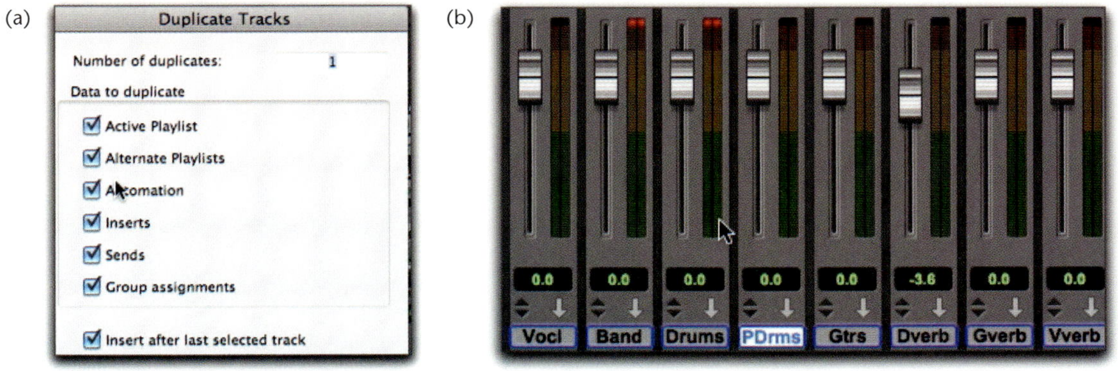

Figure 7.41 (a, b) Duplicate Your Drums Aux Track and Rename It.

"All-Buttons-In" Compression

Now let's add a BF76 compressor to the PDrums track, and use an old technique that some say was discovered by accident, known simply as "All-Buttons-In." With all of the ratio buttons pressed, many engineers refer to the compression sound simply as "Nuke mode."

Shift+click on any of the Ratio buttons, and you will see all of them press in.

Figure 7.42 All-Buttons-In Mode on the BF76.

Now, because we duplicated the Drums submix, you'll notice that it's the exact same drum mix coming into the PDrums track. To get a feel for what it's doing, pull the PDrums fader all the way down, and mute the Guitars submix track, so that we just hear the drums.

Figure 7.43 Blending in the Parallel Compressed Drums.

Adding Urgency to the Drums

Mute the Drums bus and set the PDrums fader up to unity. Then, set your BF76 Attack to 1 and Release to 7 and start to push the input level up, until you really hear the drums smacking into the compressor. At this level, the woofiness of the kick microphone is really hitting hard, and the snare drum is positively slamming.

Figure 7.44 Aggressive Parallel Compression.

Now by itself, that sound is cool, but not usable as a final track. So in keeping with the idea of parallel compression, pull your PDrums fader down, unmute the Drums submix aux track, and start to ride the PDrums fader up, until you hear the heavily compressed PDrums track, adding a strong sense of urgency to the drum performance. Mine ended up around –12.8 dB, but be sure to toggle the mute switch on the Pdrum track and set the level of your's to taste.

Figure 7.45 Blending PDrums for a Rocking Drum Sound.

The one thing I don't like in this parallel compression sound is that because all of the drums are going through the PDrum track, it's also squashing the cymbals and overheads a bit more than I like.

Multiple Track Output Trick

Although we could create a separate, postfader send from the rest of the drum tracks to a new bus that would feed the PDrums track, we can also effectively create a permanently postfader send by routing each of the drums we want heavily compressed out multiple outputs.

To do this, start by changing the input of the PDrums track to an unused bus like 13–14. Then rename the bus to PDrums.

Figure 7.46
Setting a New Bus as the Input to PDrums Submix.

Next, while pressing the Control key, click on the output selector for the Kick track and choose the PDrums bus.

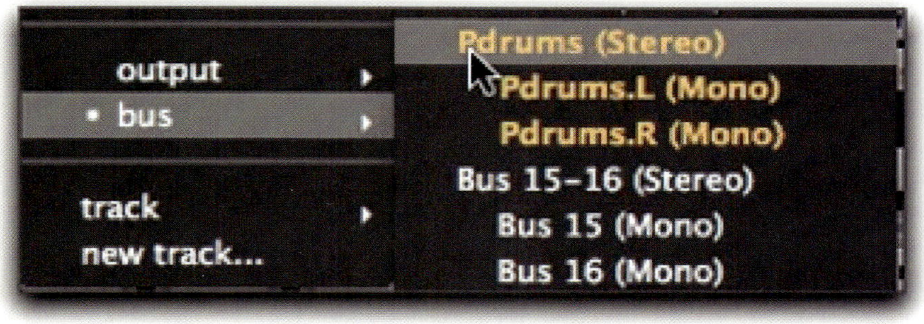

Figure 7.47 Setting Multiple Outputs for Your Kick Track.

You will now see a little "+" symbol on the output selector of the Kick track, indicating that it is going out multiple outputs (in this case, the Drum bus and PDrums bus).

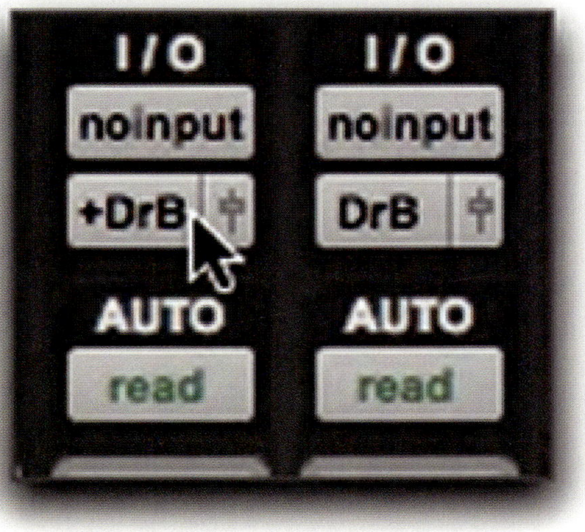

Figure 7.48 The + Symbol Indicates Multiple Outputs.

Now holding the Control key again, do the same thing with the snare and Tom tracks, routing them out both the Drum bus and PDrum bus.

Figure 7.49 All Drums by the Overheads Now Going to the Parallel Compression and the Regular Drum Submix.

Now that we don't have the cymbals getting squashed and washy in with our PDrum track, we can push its level up to get a little more aggressive with the punchy drum sounds. I pushed my PDrum level up to –9.5 dB, and the drums are really starting to rock.

And lest you think parallel compression is only a technique for pumping up drums, let's quickly do a similar setup to the bass guitar to get a touch more sustain out of it. This time, rather than routing the bass out multiple outputs, simply duplicate the bass track, name the new track something like P-Bass, add your BF76 with our standard attack and release settings, and push up the input level, until you hear the bass really starting to sustain (I hit ~–22 dB).

Figure 7.50 Compression on the Duplicated Bass Guitar.

Nice. Now solo them both out and ride the faders, until you're hearing a solid blend of the original performance with the extra sizzle of the P-Bass track.

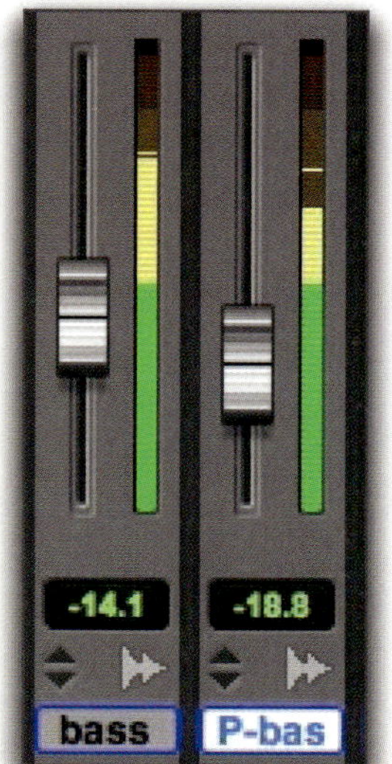

Figure 7.51 The Bass and P-Bass Sizzling at –14 and –18.8 dB.

By blending these two, we're increasing the apparent volume of the bass track. In modern rock music, where the whole track is insanely loud, it's often the result of lots of compression.

Although we could, conceivably, also use parallel compression on the guitars, it's really not necessary as we already have such a large stack of heavily compressed guitar tracks. So, it's time to finally move on to the moment of truth … the vocals.

In This Chapter

8

Vocal Mixing

This song currently only has a lead vocal and a filtered vocal for the bridge. Unmute them and let's listen to them in the track. Blend the level into the track (approximately –1.2 dB), so that you can get a feel for how it sits in the track.

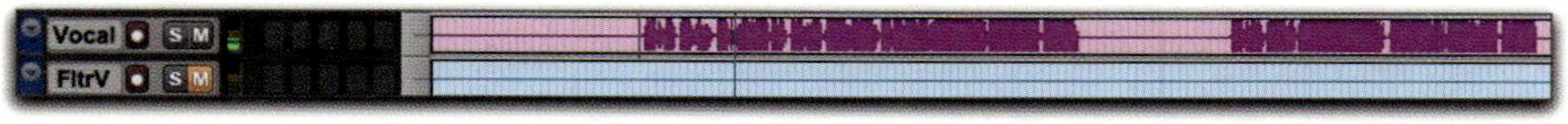

Figure 8.1 The Two Vocal Tracks.

Vocal Compression Secrets

One thing about a rock-n-roll vocal is that it often takes a *lot* of compression to make it stand up in the mix against the heavily compressed drums, bass, and guitars. So solo the vocal track now and set up our trusty old BF76 compressor with the slow ATTACK and fast RELEASE settings at a ratio of 4:1. Now when playing back the track, slowly turn up the input until your meter is showing around –1 dB of gain reduction.

Figure 8.2 Settings on My First Vocal Compressor.

© 2011 Nathan Adam. Published by Elsevier, Inc. All rights reserved.

But before we go on too far, let's set the stage for our approach to vocal compression. My technique has grown and shifted over the years, and is, currently, most like one of my favorite mixers, Chris Lord Alge. His approach is to never let any individual compressor work too hard, or drastically, on the vocal, but rather to use multiple compressors, each doing small amounts of gain reduction/compression.

What it's going to take to make a rock vocal really "pop" out of a mix is to let the attack of the notes cut through, and then compress the rest of the note more heavily. That's one of the reasons we're using a slow attack on our general compression setting. If you compress the transient at the beginning of the notes too much, it reduces the intelligibility of the sound. And if listeners (or the A&R label guy) can't understand the vocals, they will assume it's because it's too quiet, even if it's sufficiently loud already.

With that in mind, let's Alt+drag a copy of the BF76 we just made down to the next insert slot, essentially compressing the already slightly compressed vocal.

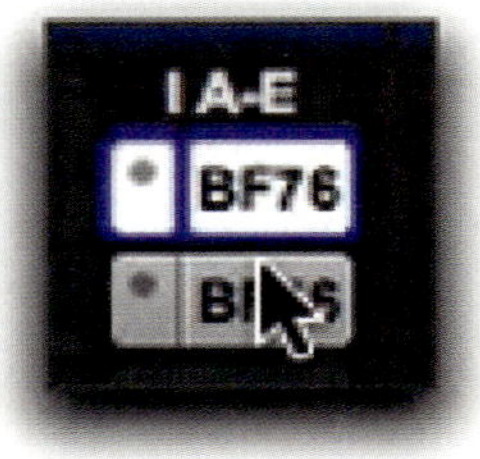

Figure 8.3 Building a Compression Chain.

Whoa! That's starting to make a difference. You'll notice that since we just made a copy of the other compressor, the second compressor is taking his "sweet" sounding voice and really starting to make it more "in your face." Believe it or not, on very rare occasions when mixing someone else's recordings, I have had as many as four to five very gentle compressors each performing slightly different amounts of compression on one vocal. Thank God for automatic delay compensation!

Vocal EQ Technique

Before we get too far down the road though, with the compression on the vocal now, I'm hearing some deficiencies in the sonics of the vocal that need to be addressed with EQ. So drag your two compressors down one insert level and put an EQ7 in the first insert slot on the vocal track.

Figure 8.4
Adding EQ to the Vocal Processing Chain.

EQ or Compress

Although there are a lot of opinions about whether to EQ or compress first in a chain of processing, my general approach is to EQ first if the EQ is being used to subtract elements out of the sound (like bass buildup or woofiness in the low mids). If you think about it, in those situations, we're using EQ to make it sound like we wished it had sounded at the microphone. Then, with the frequency junk no longer contributing to the amplitude of the signal, we're able to set the compressor with more gentle settings, and get a more natural sound.

Because we've already done this process several times, you know the procedure. Start by filtering the low end with the HPF. My settings ended up at 18 dB/octave at 206 Hz. His vocals feel a little lifeless in the high end, so let's add some top with the HMF EQ. Again, start with a medium *Q* of 1 or so and sweep the frequency around, until you hear the frequency that helps this vocal pop out in the mix.

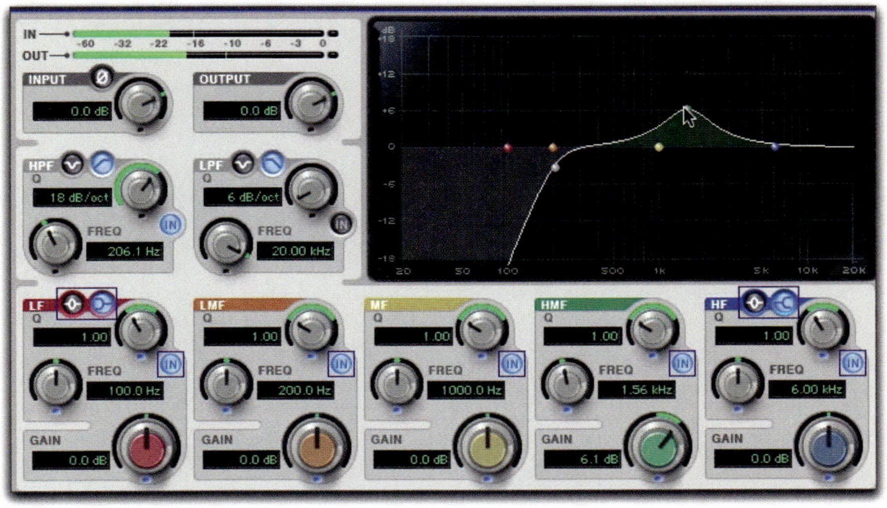

Figure 8.5
Searching for the Right High End to Add to the Vocal.

Then, broaden your bandwidth (or *Q*) and lower the gain until you've got a nice, natural-sounding bump in the highs, without sounding overprocessed.

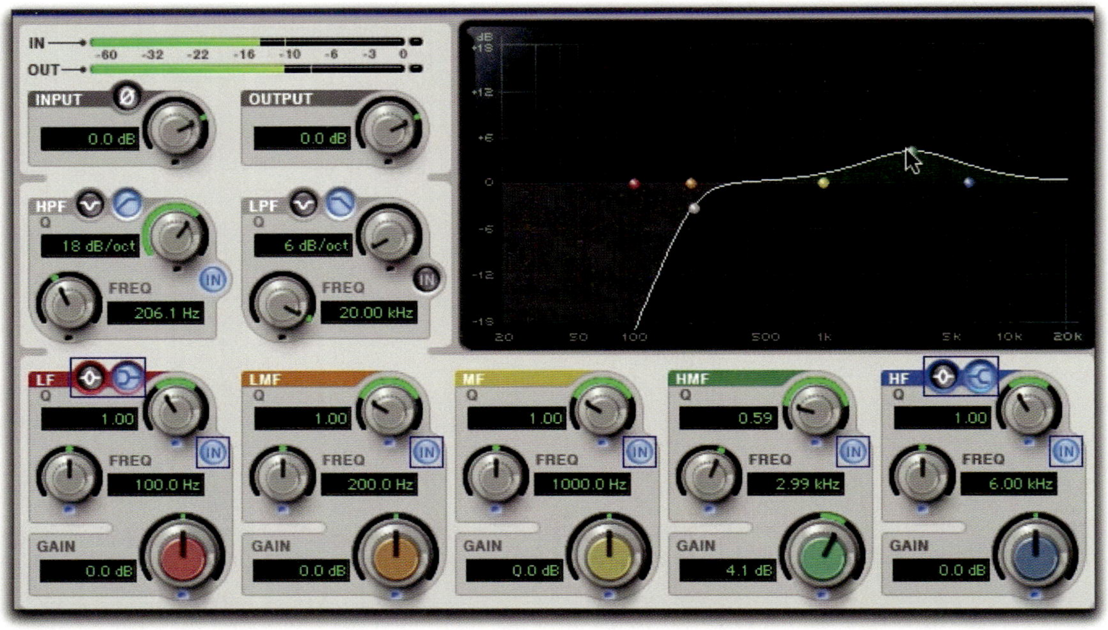

Figure 8.6 That's Better.

Finally, unsolo his vocal and blend the fader level with the rest of the music. Mine ends up around –6.8 dB.

Vocal Delays

With the vocal level roughed in, let's look at his reverb and delays for a minute and explore adding a rock-styled delay to his reverb return. To start, we need to create a send from the vocal track out to the Vverb bus.

Jump over to the Vverb track, move the reverb plug-in to the second slot and bypass it, then add a plug-in>Delay>Extra Long Delay II (mono) plug-in into the first insert slot. Because the reverb plug-in we used was mono to stereo, this mono delay will be all we need.

Figure 8.7 Creating a Send from the Vocal Track to the Vocal Verb Bus.

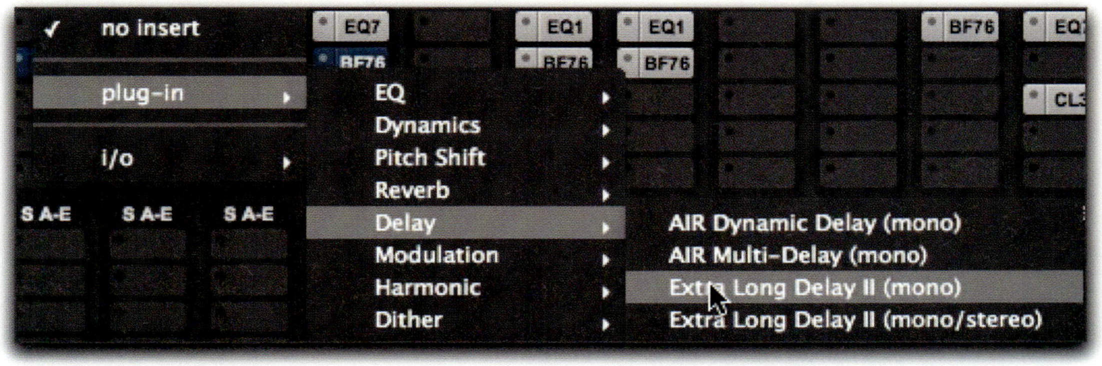

Figure 8.8 Adding a Long Delay to the Vocal Reverb Track.

There are some great presets in this delay plug-in, and as always, we can use the plus and minus buttons to toggle through the presets to find a starting point while listening to the vocal level in the track. For a song with a medium rock tempo like this, I like to start with a quarter-note delay, and then add eighth note delays if the song needs some more lively elements. After toggling through the list, I ended up with one called 4th Delay (Long Dly).

Verb versus Delay

One of the reasons I like to start with delay on a rock vocal (rather than reverb) is because reverb tends to sweeten sounds (e.g., strings) and push them away from the listener ... which is not really what we're going for with an upfront rock vocal. Delay, however, doesn't move your singer back on the sound stage, and yet it feels reminiscent of the sound and movement that reverb adds to a sound.

The first thing I do on a delay plug-in is bringing the Mix level way down as it's rare for a track to need giant, loud delays. The Mix setting on virtually every plug-in goes from 0% wet (or, no plug-in sound) to 100% wet (fully delayed signal at the output). Set it around 11% to start, and we'll blend more as needed.

Figure 8.9 Setting Up My Vocal Delay.

It's sounding nice with those settings, but the echos are a little too present with almost the full frequency range of the original signal. Similar to how we don't want the listener to notice vocal tuning, we would rather them *feel* the effects of the delays, rather than throw the delay sound right in their face.

Filtering Delays

To accomplish this, grab the low-pass filter (LPF) and start to pull it down, effectively rolling off the high end of the delayed signal, and making the delays more *felt* than *heard.* My LPF got all the way down to 7600 Hz before I felt like we'd pulled enough of the wispiness out of the delays. Now that the sizzle is less prominent on them though, I think we can pull the mix of the delay back up to feel it a bit more in the mix. Try 14% and see how it feels to you.

Figure 8.10 A Good Feeling Vocal Delay.

Now obviously, depending on the budget of the record, you can (and I have) spend an entire day trying different effects and tweaks. Given the state of modern recording budgets though, where entire records are often mixed in a matter of days (or less), we've dialed in a great-sounding delay that brings a lot to the track in only a few minutes because I have created and collected presets that, I know, work for various styles of music over the years.

Make Your Own Presets

That's why it's so important that you spend some free time developing and saving your own presets as you work on each new record because they'll give you a quick springboard to get back to "that great delay sound" you had once upon a time.

Moving on, this delay feels so good that I'm not sure we're even going to bring the reverb plug-in back to the track. So in the mean time, let's move on to automating the vocal ride.

Even with the ubiquity of modern compression plug-ins and techniques, a good vocal ride can still be the difference between a dull vocal track, and one that rides blissfully above the music, soulfully inviting the listener on a musical journey.

Vocal Riding with Automation

Alright, that might be a bit overstating it, but we still need to make sure this vocal doesn't get lost in the mix of raging guitars and drums. To accomplish this, make sure the AIR plug-in is still summing our mix to mono, as it will ensure that if our vocal rides above the track in mono, it will maintain a much better balance against the track in stereo.

So let's set up the track to do an initial vocal ride to the entire vocal. The first step (that is often neglected by amateur engineers) is to open the Automation window (Mac: Cmd+4, Win: Ctrl+4) and disarm all of the parameters you *don't* want to automate right now. By default, Pro Tools has all of the parameters armed, so go ahead and disable all but the volume switch. I'd like to be able to tweak plug-ins, pans and sends up to the last second of the mix, so I'd rather not have to constantly fight old automation that I didn't mean to record in the first place.

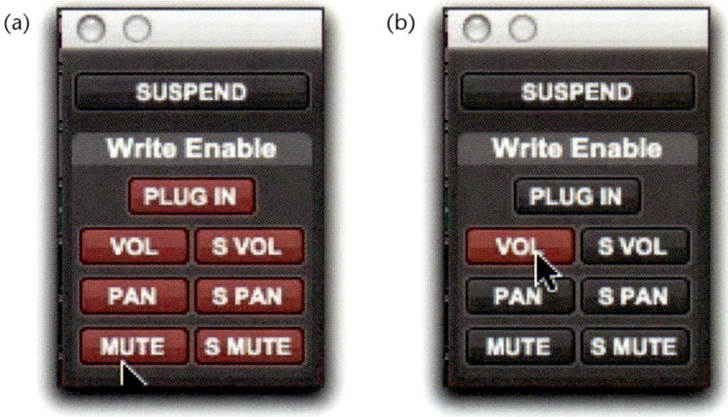

Figure 8.11
(a, b) Disable All but the Volume Automation.

Now switch your Vocal track view to Volume, and set its Automation mode to touch, and let's begin to ride the vocal level over the course of the track.

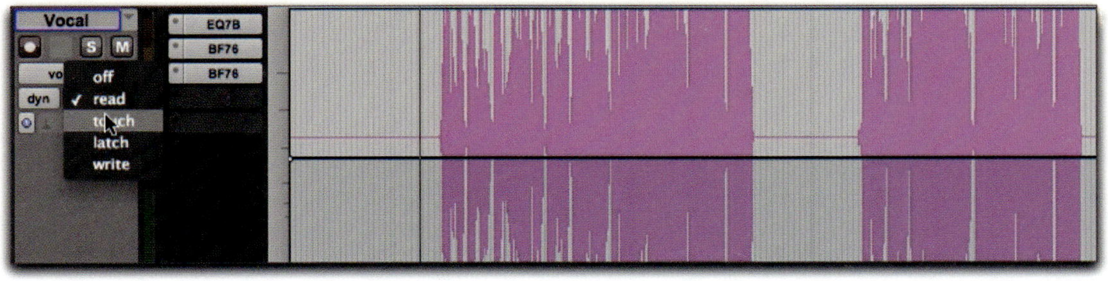

Figure 8.12 Automating the Vocal Volume.

On the first pass through, you want to listen to find an initial level that's not going to require constant, giant jumps of the fader. During my first pass, you can see by the red automation line that my initial level started out a bit too hot, so I started to pull it down by around bar 15.

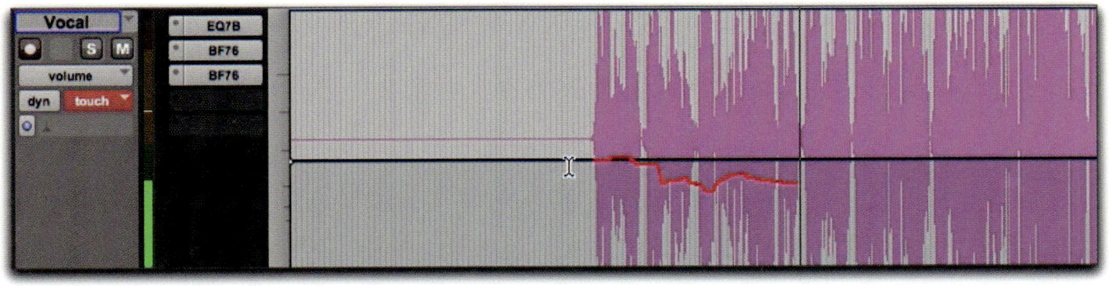

Figure 8.13 The Beginning of the Volume Ride.

Tweaking Your Vocal Ride

Rather than spend the entire track trying to ride the volume level lower, let's quickly open the second BF76 compressor and pull its output level down to around –22 dB.

Figure 8.14 Reducing the Overall Level of the Second Vocal Compressor.

Now, after looking at the fader level we're ending up with as the song plays on for a minute, I'm noticing the average level is around –8.9 dB on the fader. So let's select and delete any automation that we've already made, and pull the entire Volume envelope up so that it starts at –8.9 dB.

Figure 8.15 (a–c) Finding and Resetting the Initial Vocal Level for the Vocal Ride.

Remember, what we're listening for is the ability to clearly understand the vocals, phrasing, and emotional intensity of the singer. Also, because we took the time to find a good initial vocal level, you'll notice that the fader moves I made over the majority of the vocal track are just pushing up lines, words, and phrases that either added emotional intensity to the performance, or were in danger of getting lost in the mix. Go ahead and compare your vocal ride against the vocal automation in the Pro Tools session file.

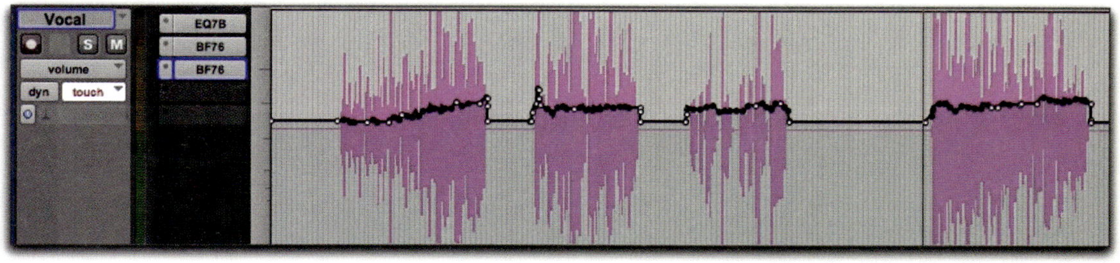

Figure 8.16 The Vocal Ride.

Now if you look closely at that automation, you'll notice that over the course of the track, I was riding hotter and louder. Although there was less instrumental competition for the vocals at the beginning (hence, less need for a hotter vocal), the beauty of nonlinear editing is the ability to quickly jump between the beginning and end to see if the vocal has increased in volume disproportionately to the music.

So, quickly jump through the four main vocal sections and you'll see that if your ride went anything like mine, while I do like the actual moves we made along the way, you will definitely get significantly louder as you go.

The ideal would be if we could keep all of our automation rides exactly as they are, and just trim the whole level down a few decibels along the way. Of course, this is a function available in HD, but because many more people have regular Pro Tools 9, let's look at a few other options.

One, you could automate the level of the actual Vocal Aux submix track that the lead vocal is ultimately going out of. However, what if we later added some background vocals to that bus? Well, they'd be going through the same automation.

Another approach would be to just grab the entire automation envelope with the selection tool and trim the points up or down.

Figure 8.17 The Lazy Way of Trimming All the Automation Up.

However, let's take the nonlazy approach of listening to the various parts of the song and deciding whether their intro needs to come *up*, or the outro needs to come *down*.

Level Comparing with Pre- and Post-Roll

First, let's open our Transport window (Mac: Cmd+1, Win: Ctrl+1) and set 1 bar of Pre-roll and Post-roll, so that as we listen to the track, we'll always have a few seconds of musical context for the section we're now attempting to work on. You can always quickly engage or disengage the Pre- and Post-roll with the Cmd+K (Win: Ctrl+K) shortcut.

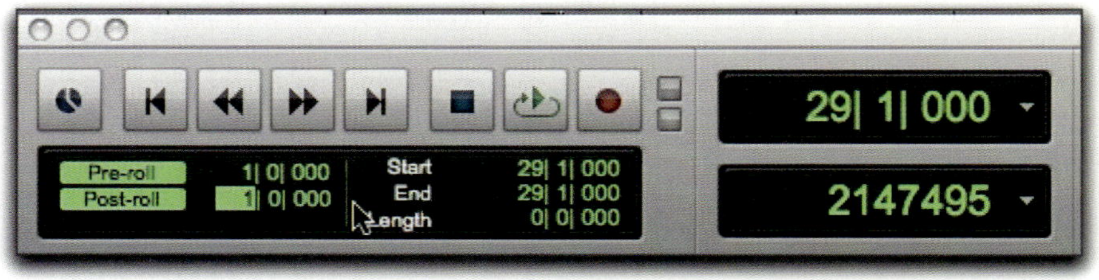

Figure 8.18 Setting Pre- and Post-Roll for Providing Some Mix Context.

Now, by opening the Memory Locations window (Mac: Cmd+5, Win: Ctrl+5), we can quickly jump between the various choruses to compare the levels of the vocals in each one. Not only that, but because we have Pre-roll turned on, we'll also get an extra bar of music before each chorus to ensure the levels before the choruses are transitioning similarly as well.

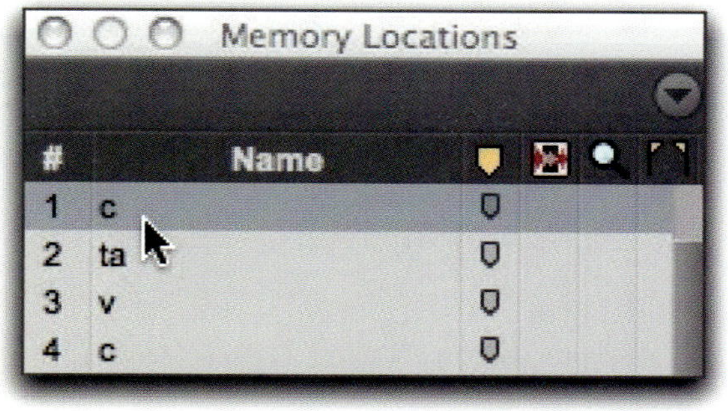

Figure 8.19 Memory Locations Allow Us to Quickly Toggle between the Various Parts of the Song.

Jump between Memory Locations 1, 4, and 10, and you'll quickly hear the levels of the various choruses.

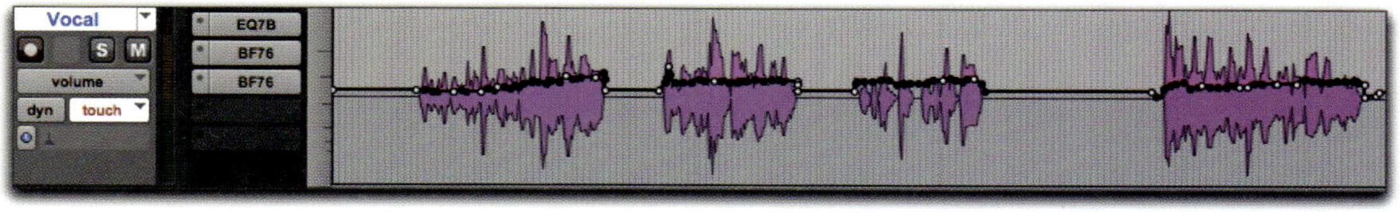

Figure 8.20 Comparing Vocal Levels across the Song.

After comparing them, it sounds to me like only the last chorus started to get a bit out of the pocket volume-wise. So let's select the last chorus and start to trim it down. As a rule of thumb, half a decibel can be felt, while ¾ of a decibel is where you start to hear the difference. So let's start by selecting all the automation over this last section and pulling it down around ¾ of a decibel.

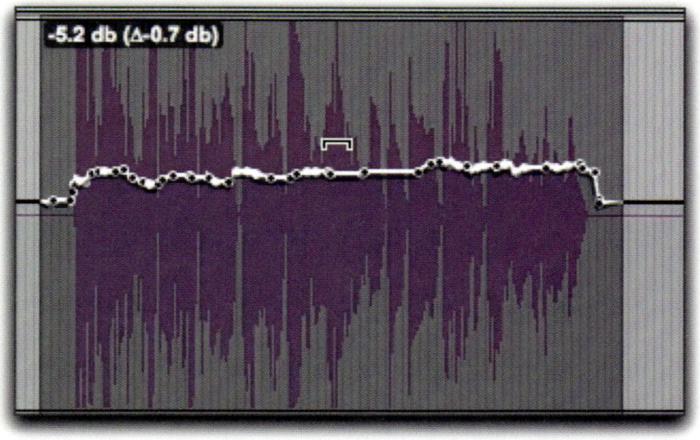

Figure 8.21 The Numbers on the Left Show the Current Level of the Track. The Numbers in Parenthesis Show the Change from the Existing Automation Level.

Micro-Tweaking Level Trick

If you just click and drag the automation level down, you may notice it jumps too fast to tweak fractions of a decibel. Just hold down the Command key (Win: Ctrl) and it will now adjust the volume in tenths of a decibel.

Continue to reference between the choruses, and trim your levels as needed. I ended up pulling the last chorus down another –0.5 dB. Be sure to reference your vocal ride against the one in the Pro Tools session, but I think we're ready to look at that filtered vocal.

Start by copying the plug-ins from the lead vocal down to the filtered vocal.

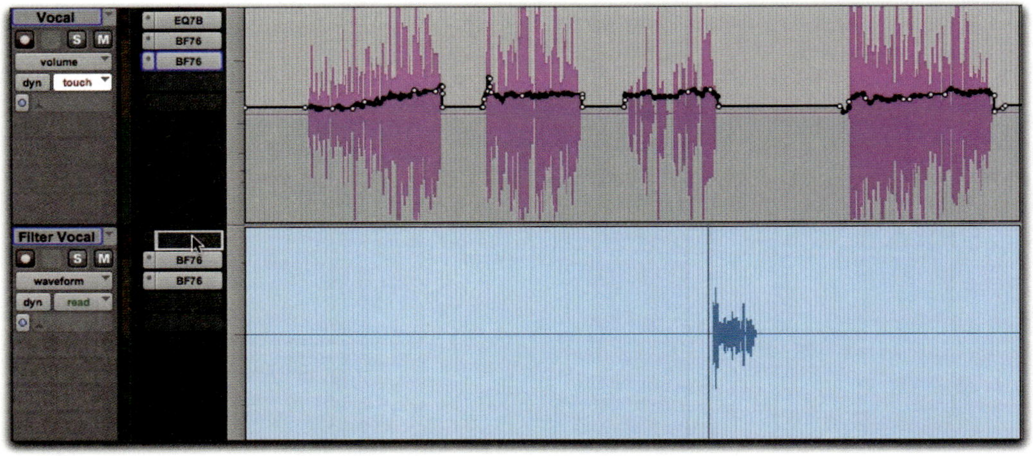

Figure 8.22 Copying Plug-Ins Is a Good Starting Point for the Filtered Vocal.

Now mute the lead vocal for a second, and let's concentrate on getting that classic, band-limited telephonic effect on this track.

Vocal EQ Effects

Because this track is more whispered than sung, let's start by adding a touch more gain (2–3 dB should do it). Open the Filtered Vocals EQ and turn up the input knob about 2.7 dB. Now, let's raise the Frequency of the HPF, until we hear the vocal really starting to thin out (try up near 439 Hz). Now, let's boost the existing HMF, until we really feel the sizzling on the vocal track. I like the funky, almost hornlike effect we get with an 8.6-dB boost at 1.53 kHz. Finally, let's kick in the low-pass filter with an extreme slope of 24 dB/octave. Pull it down until it's almost too dark, and then ease it back up a notch. Finally, adjust your HMF until it sounds sufficiently "telephonish" and compare it to the EQ curve we arrived on the next page.

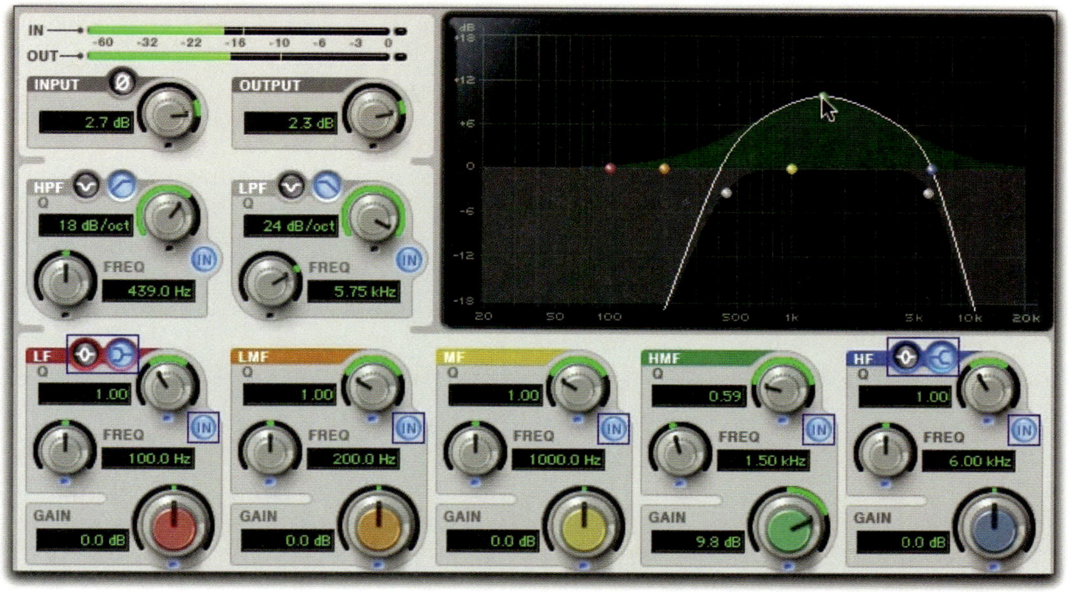

Figure 8.23 A Seriously Honky Telephonic Sound.

To give it even a little more spank, let's hit it hard with the compressors again to see if we can really get that squashed "AM Radio" sound. Crank the input of the first one up a couple of decibels, with a similar nudge up on the input of the second compressor. Finally, adjust the output gain (mine is around –24 dB) until the level of the vocal sits properly in the track.

Figure 8.24 Squashing the Filtered Vocal.

Groovy! That sounds nice and smashed. Let's do a quick vocal automation on it and we'll be about ready to move on.

Drop the filtered vocal into Touch mode and listen through to it. Make a pass if you like, but after a couple runs through, I'm feeling like the only line getting lost between the drums and guitars is the first "40 days and …". Let's do a quick volume bump at that point, fading back to the existing level at the end of the phrase.

Figure 8.25 Just a Little Bump at the Top of the Filtered Vocal Part.

While we're at it, Alt+drag the Vverb send from the vocal track over to the Filtered Vocal track and let's add a touch of that sweet delay to it.

Figure 8.26 Copying the Vocal Verb Delay Send.

Cool! I feel like that's working really well. So the last thing we need to do for this part is look closely at the transition between the regular vocal and the filtered vocal at bar 85.

Oops! Because we were soloing out the filtered vocal, we didn't notice how much louder it is than the lead vocal track as we were automating it. Select the whole chunk of automation over the Filter Vocal section and let's trim it down to 5 dB. Give it another listen through.

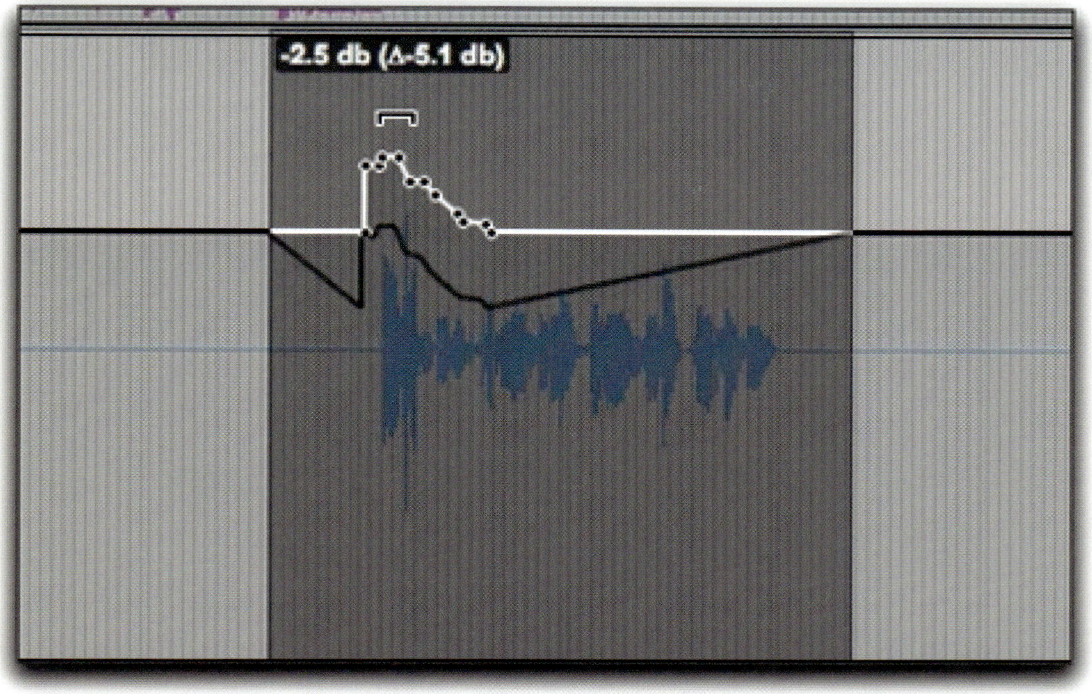

Figure 8.27 Trimming the Filter Vocal Down to 5 dB.

Ear Referencing

There's a valuable lesson to be learned about mixing here, and it goes something like this: Your ear is a drum. It changes the way it hears things day to day and even minute to minute, depending on how hard it's been working, the weather, the air, how you feel, and so on. This is why even professional engineers, when working in a new studio, will pull up other commercial reference CDs that they know by heart when they are learning the particular sonics of a new studio, monitors, and so on. Without having an unchanging reference like this, it's impossible to know that your ears aren't tricking you into making poor mix decisions.

Well, since this is a book, I could have just eliminated that last paragraph because on listening at the new lower level, I'm actually thinking we should just delete the automation on the filtered vocal, and set the whole track to –5.9 dB. Just like we didn't feel compelled to use the drums room microphone that didn't complement the track, we don't need to use automation just because it's there.

When to Not Automate

The only reason I'm keeping that last series of changes in is so you know that not every decision you make during the mix process is going to be, or should be, kept. There are times in mixing that you will get an hour into setting levels and effects, only to come to terms with the fact that the approach you were attempting just isn't going to work, and you'll start over.

Ideally, as you gain more experience, this will happen less and less, but it's still a part of every mix to have to set levels, EQ's, effects, and then as tracks are added, return to these same things and make changes. C'est la vie.

Beautiful! I like where the vocal tracks are placed in the song, so I think we're about ready to pull into the home stretch of mixing this track.

In This Chapter

9

Timing Fixes

Well, we've spent a lot of time building a great mix of drums, bass, guitars, and vocals, and as we get farther into the mix, we get to make, increasingly, tweaky mix decisions. First, while I like the general level of the acoustic and electric guitars that play through the whole track, let's dig into the acoustic guitar track and make sure we've got the optimum base level for it.

If you want to reward yourself on how far your mixing has really come, you can also toggle the AIR Stereo Width plug-in off for the first pass so that you can hear how well your balances are maintained, all from forcing yourself to build the mix in mono. Great job!

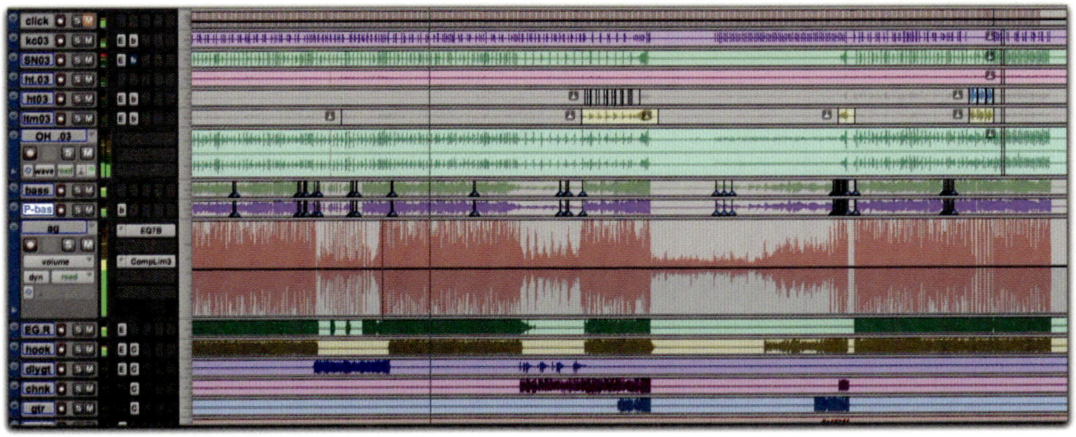

Figure 9.1 Listening through the Track in Stereo.

Once you've done that, play through the track and, periodically, toggle the solo button on the acoustic to hear how it's fitting. From what I hear, the level is balancing really well with the exception of the chorus breakdown (Memory Location = c bd), where it feels kind of low. Let's put it in Touch automation mode and fix that, shall we?

© 2011 Nathan Adam. Published by Elsevier, Inc. All rights reserved.

Acoustic Guitar Automation Tweaks

To find the level you're going to want, role into that section and push the fader up until it supports the vocal without overpowering it. Once you've found a good level, you can stop playback and click on any of the automation points to get readout on what level you should be shooting for. In this case, it looks like –0.7 dB.

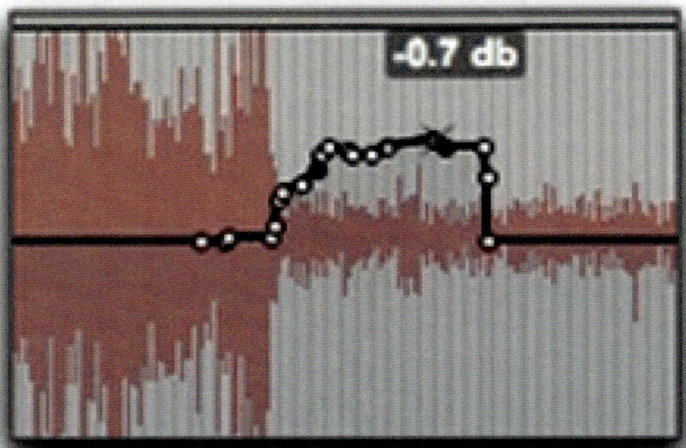

Figure 9.2 Finding the Right Level for the Acoustic in the Chorus Breakdown.

So delete the automation you just wrote, and let's try just selecting and trimming this entire section up to –0.7 dB.

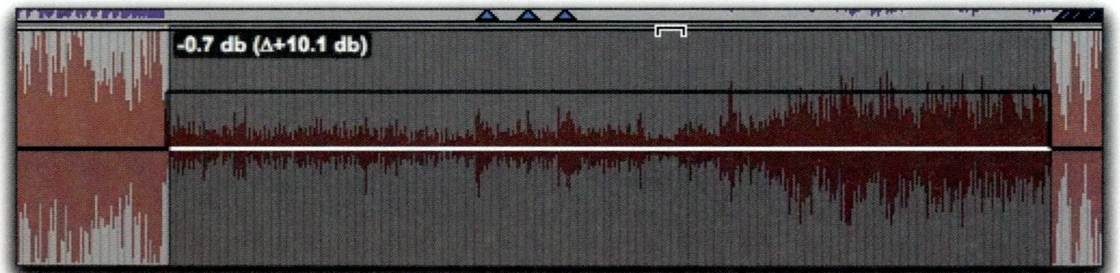

Figure 9.3 Quickly Trimming the Breakdown Section Up to –0.7 dB.

The only problem with automating this way is that it's very easy to either miss notes at the beginning or at the end, which you then have to manually correct for. I did exactly this by trimming after the first acoustic note of the breakdown, which has the impact of making the guitar seem too quiet overall.

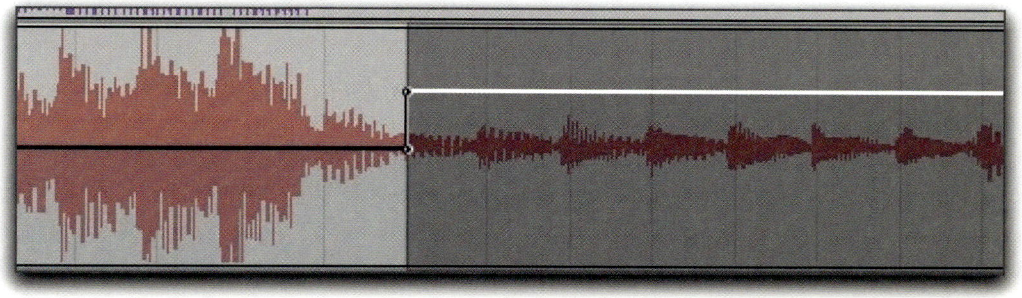

Figure 9.4 Oops! Missed That First Acoustic Break Note.

Simply use your Smart tool or Pencil tool to go in and build enough of a ramp over that note to utilize what we learned about back during drum compression: If the first note or part of a note is loud enough, it grabs our attention and we tend to perceive the whole part as loud enough. If the first part is too quiet, we'll tend to perceive the entire part the same way.

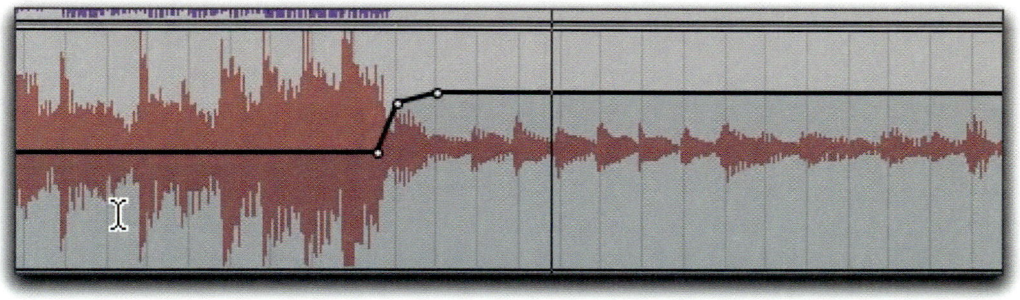

Figure 9.5 Ramping Up Over the First Note.

Alright! Now that it's coming in strong enough, let's get back in Touch mode and start automating a fade back to the lower level from about bar 95, where the fugue guitars start building in. While I'm riding it, I leave a little extra in right at the last two bars of slamming guitars.

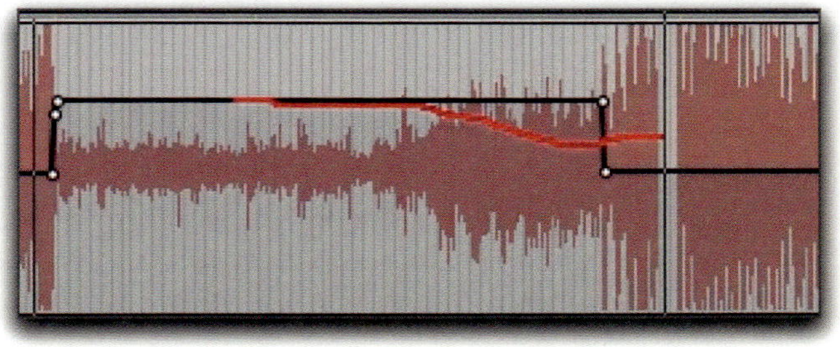

Figure 9.6 Automating the End of the Acoustic Breakdown.

We're close, but on listening with the rest of the guitars, I think we should make the acoustic fade a little lower a little earlier. As always, reference your automation against the included Pro Tools sessions if you like.

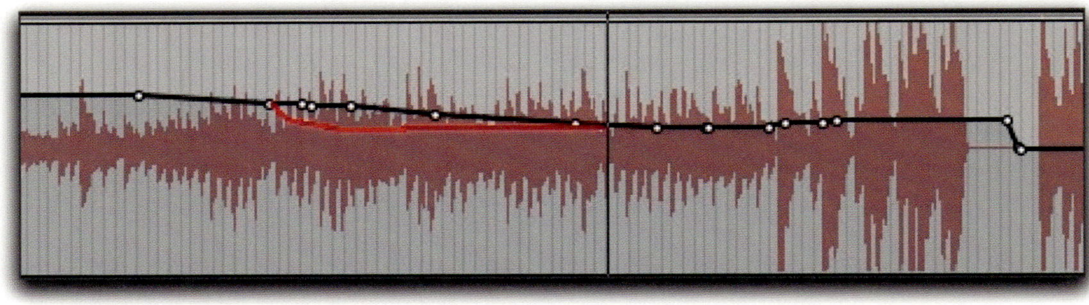

Figure 9.7 Trimming the Acoustic Fade a Little Lower.

Now that we've got the drums and acoustic feeling good, I hear a particularly out-of-pocket timing problem in the acoustic guitars. We'll take care of it in just a minute, but before we do make one more pass over the electric guitars to ensure their levels are blending well together, and that anyone doing a special part get's a moment to shine through in the mix. If they're just strumming with the group, focus on blending rather than pushing them up.

Mixing Electric Guitars

For example, let's rearrange our Chunky, End Guitar, and Guitar tracks, so that we can see them visually ramping up before the final chorus.

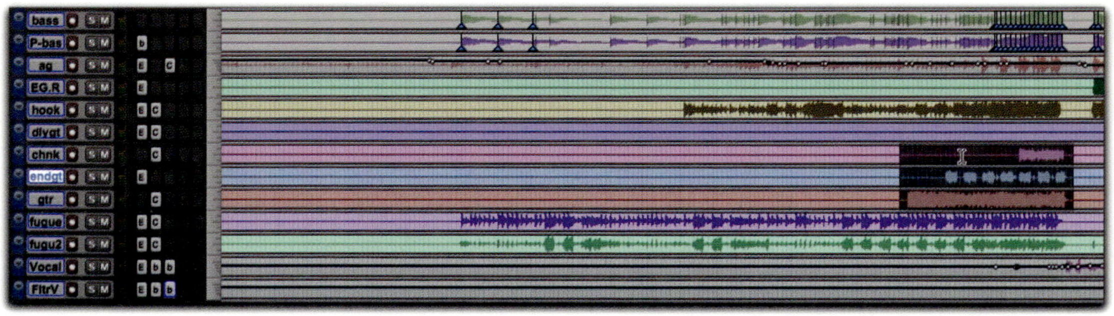

Figure 9.8 The Three Part Guitars of the Build.

Now mute the two that start later and make a pass on the Guitar track. Then mute Guitar and make a level pass on the End Guitar, followed by the same procedure for the Chunky guitar. It's like musical chairs mixing. My levels end up at Chunky –15.8 dB, End Guitar at –9.6 dB, and Guitar at –17.2 dB.

Figure 9.9 The Blend for Your Extra Guitars.

Well, the toughest part about mixing them is hearing the massive timing problem that is happening somewhere in the final buildup. They really clash and will likely need a combination of manual pocketing and Elastic Audio to put them in place. Fortunately, we've already added these skills to our mixing toolkit, and just need to zoom in and identify the source of the problem.

Start by muting End and Chunky guitars and soloing the Guitar, Kick, and Snare tracks. This will allow us to quickly and audibly find which guitar is the source of our timing problem.

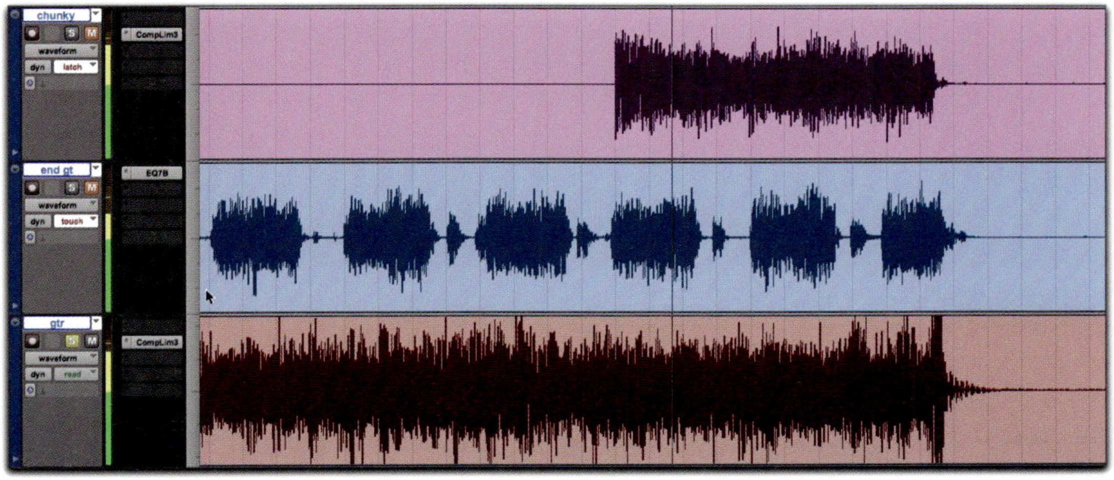

Figure 9.10 Soloing the Guitars against the Drum Tracks to Find the Timing Problems.

Pocketing Electric Guitars

After listening through, the Guitar track sounds ok, but the End guitar is felt a little late on several pieces through this section. Go ahead and turn off Elastic Audio on the End Guitar track and manually cut and nudge the first section (at 107|1), until it lines up with the beginning of the bar.

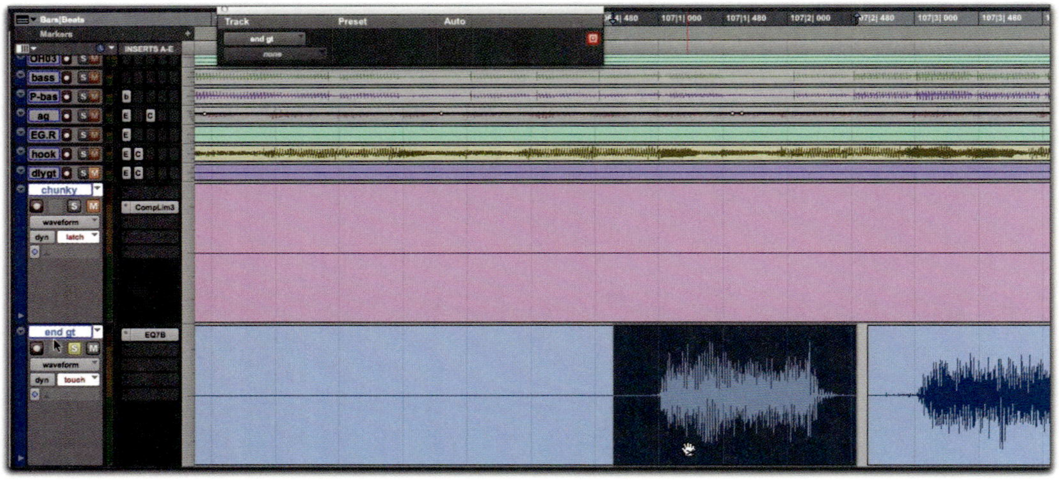

Figure 9.11 Manually Pocketing the End Guitar Track.

Now repeat this select–separate–nudge process for all of these strums, lining up the front of each strum with the grid.

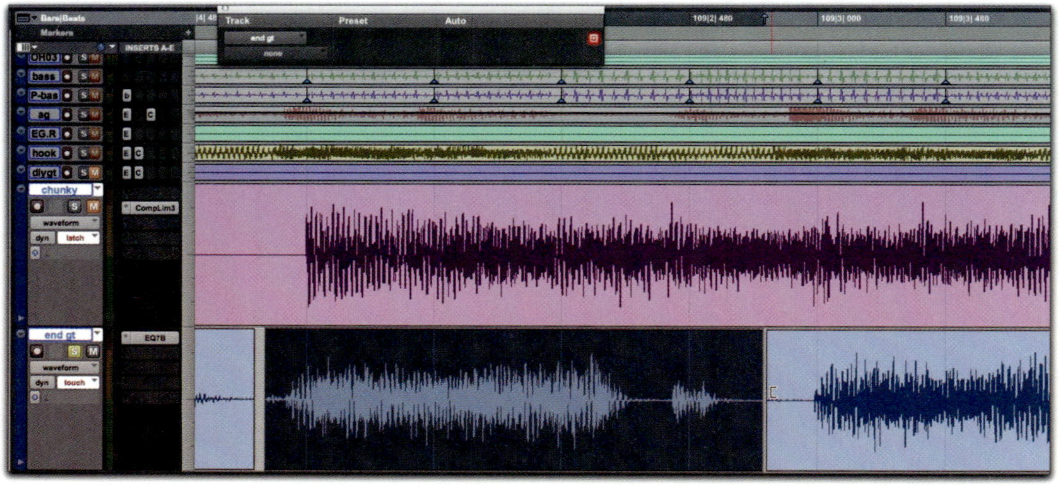

Figure 9.12 Fixing More Strum Timing.

Although this improves things, I'm not hearing any problems with the Chunky guitar track, which means the timing problem is one of the other guitars … probably the fugue guitars. Solo them out against the Kick and snare and listen to the fugues.

Pocketing Fugue Guitars with Elastic Audio

Sure enough, they're definitely having problems. Because we haven't really edited them so far, go ahead and delete any excess noise or empty areas and let's look closely to see where these fugue guitars get really off the beat.

(a)

(b)

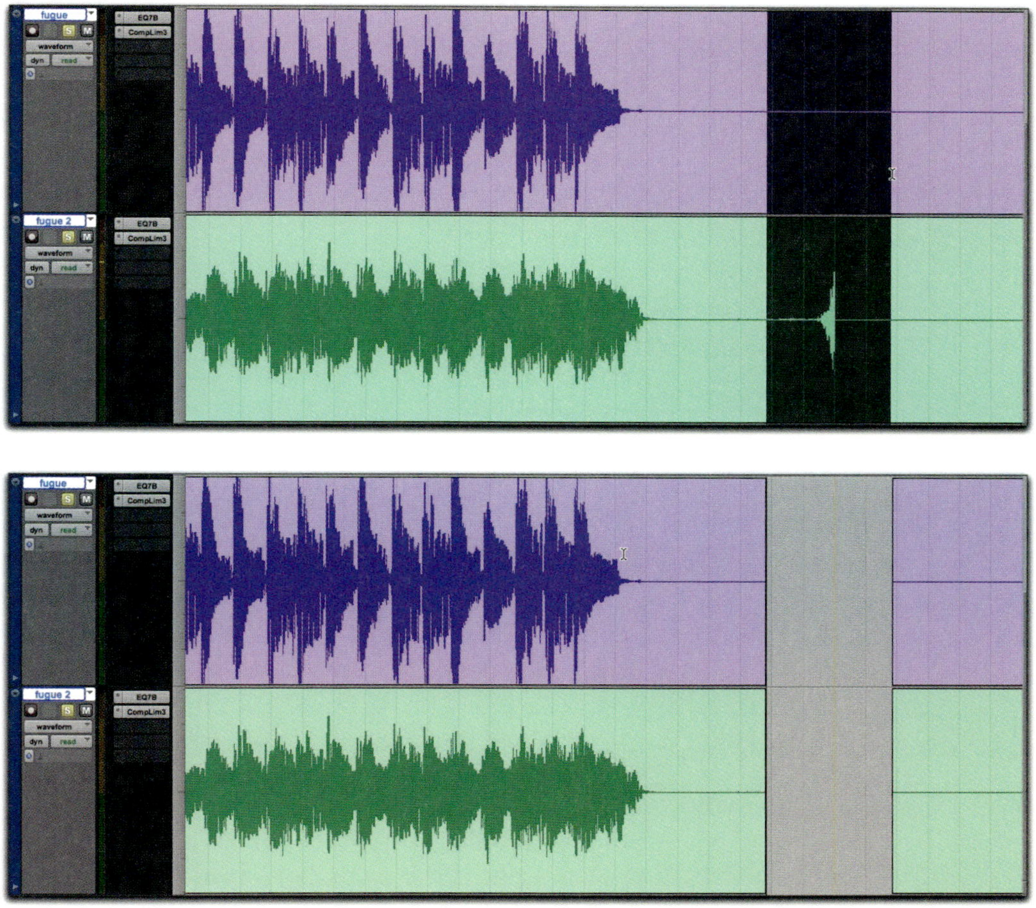

Figure 9.13 (a, b) Cleaning Up Fugue Guitars.

Now set them both to Polyphonic Elastic Audio mode, allow them to be analyzed, then mute the second fugue guitar and switch to Grid mode. This will let us rapidly pocket the choppy fugue parts against the drums. As we saw way back at the beginning of the book, just hold down the Shift key and start snapping the transients of the notes against the eighth notes.

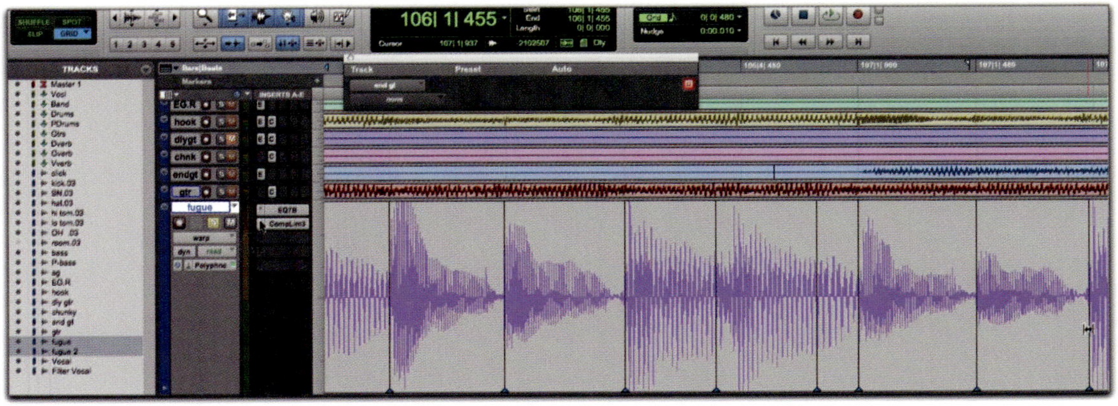

Figure 9.14 Using Elastic Audio to Pocket the Fugue Guitars.

For the last bar or so of the fugue guitar, the transients are so well defined. I saved a few seconds by simply bringing up the Event Operations window (Alt+3) and quantizing the warp markers to the eighth notes. Nice!

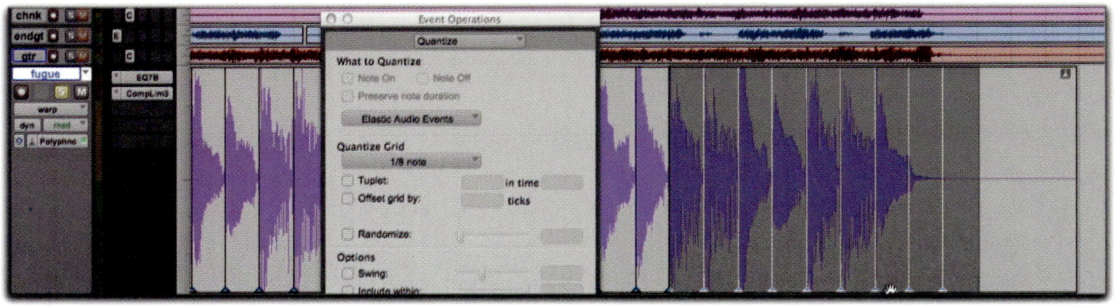

Figure 9.15 Quantizing the Fugue Guitar.

After listening back, it all sounds good except for one note. Around 106|4, the analysis has, incorrectly, put a marker visibly ahead of the actual transient.

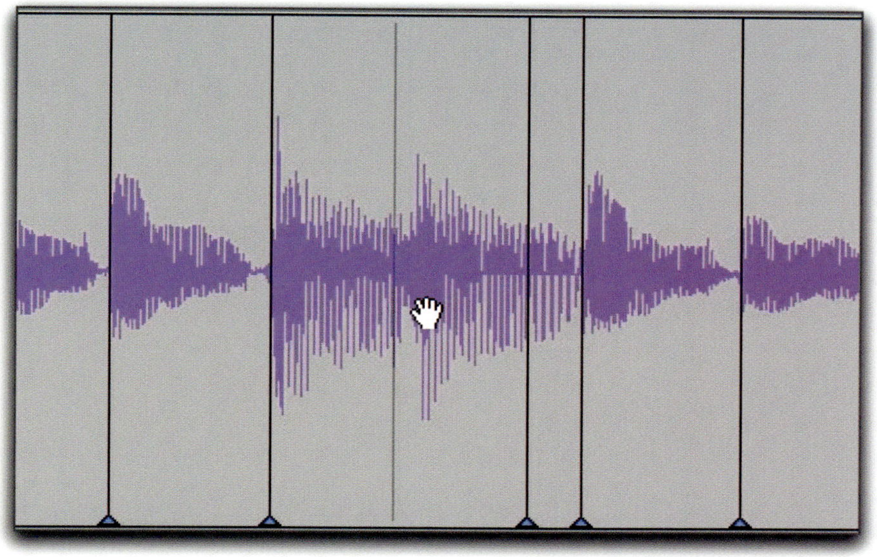

Figure 9.16
An Early Warp Marker.

When this happens, as we saw earlier, simply switch to Analysis mode and move the marker onto the audio transient. Then, switch back to Warp mode and pocket the transient into place against the grid. Above all, trust your ears.

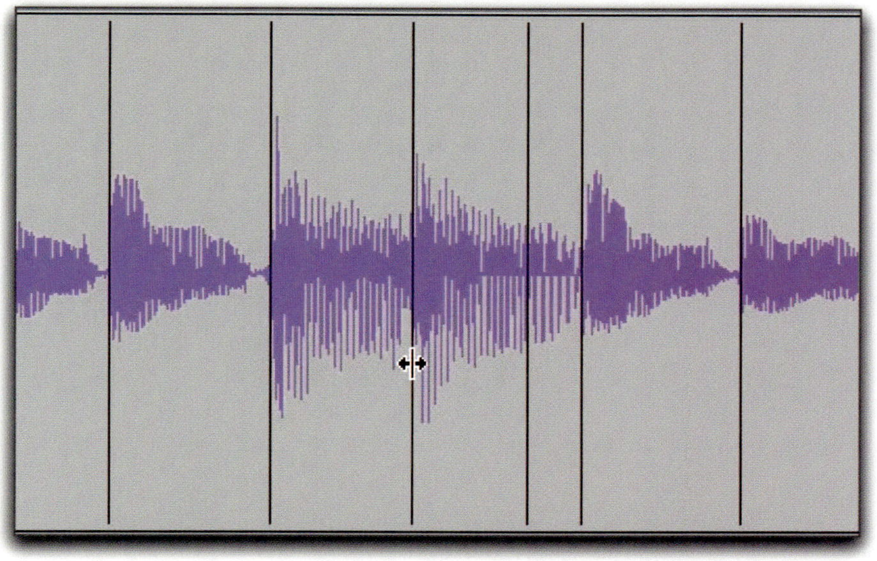

Figure 9.17
Moving an Errant Warp Marker.

Quantizing Chunky Guitars

Now that fugue guitar is in place, mute it and solo the Fugue 2 against the drums. From first listen, it sounds like he started getting ahead around the same time as the first fugue guitar, say bar 105. Let's go a little wild and try to just quantize the whole thing in one fell swoop. It rarely works, perfectly, for every note on the first try, but with a straightforward rhythm pattern like this, it's worth a shot.

Hold down Shift and drop a set of boundary markers at 105|3. Then jump over near 110 and hold the Control key to set another boundary.

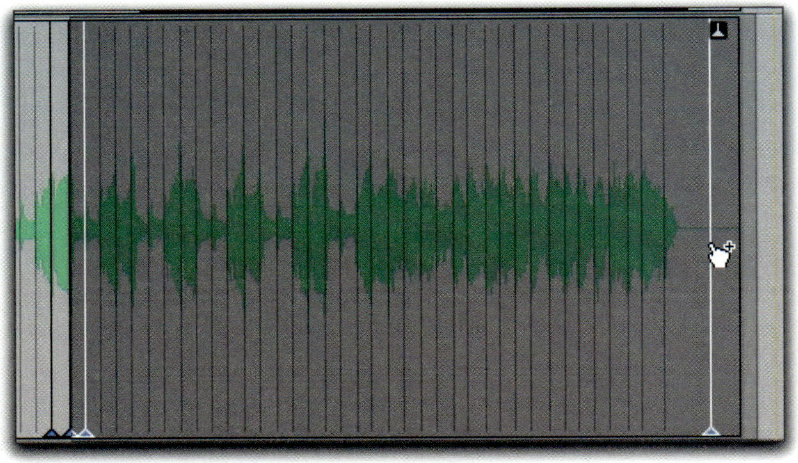

Figure 9.18 Setting Boundaries around the Area to Be Quantized.

Fire up the Event window again and quantize them to the eighth notes.

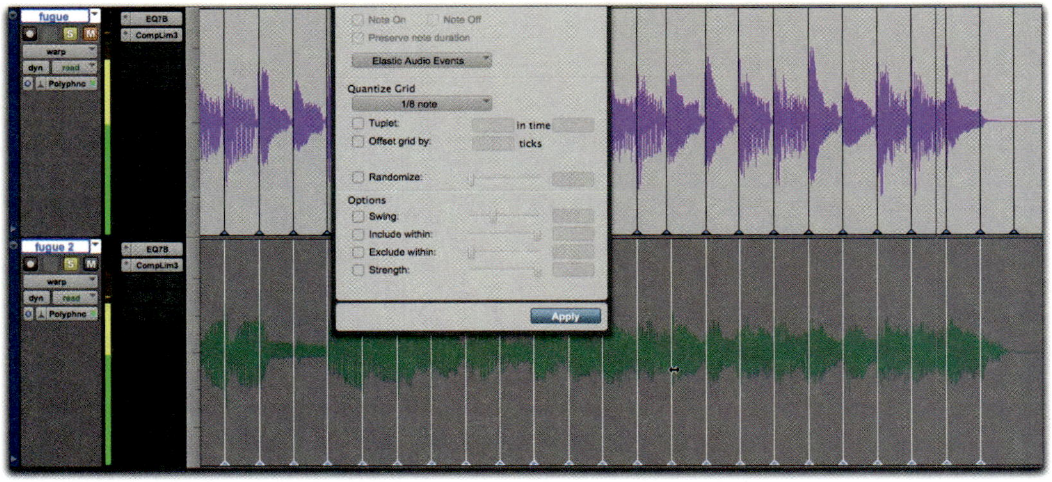

Figure 9.19 The Quantized Fugue Guitars.

I'll be darned. It worked perfectly the first time. That, my friends, is why we constantly add to our mixing toolkit, creating better mixes in less time.

Fixing Acoustic Timing

Unfortunately, that still doesn't fix the primary timing problem we're hearing, and on further investigation, I see it's the acoustic guitar. In fact, when we zoom in, we can see that the whole acoustic part is coming in early. Use the tools you've been developing to nudge the whole section into time, and then use Elastic Audio to get the guitar strums right in the pocket.

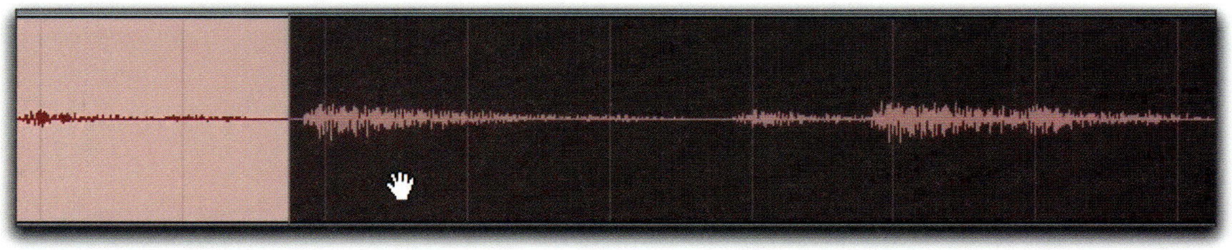

Figure 9.20 The Acoustic Is Consistently Early.

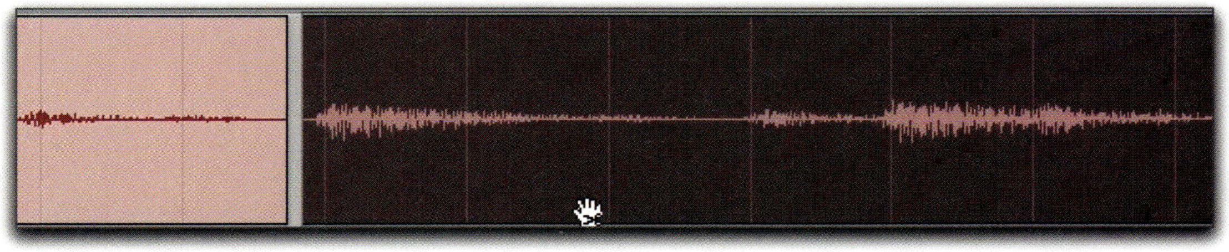

Figure 9.21 Cut the Region and Nudge It Back 20–40 ms, Until It's Closer to in Time.

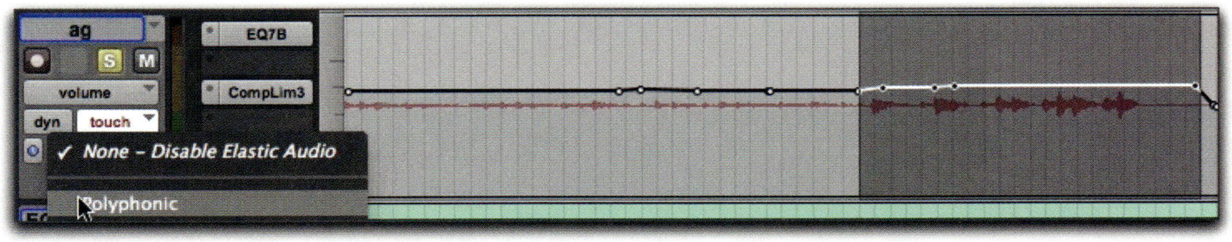

Figure 9.22 Set the Acoustic to Polyphonic Mode.

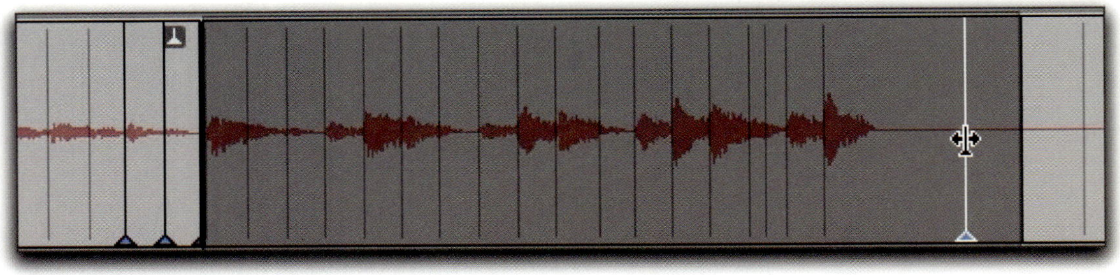

Figure 9.23 Set Your Boundary Anchors.

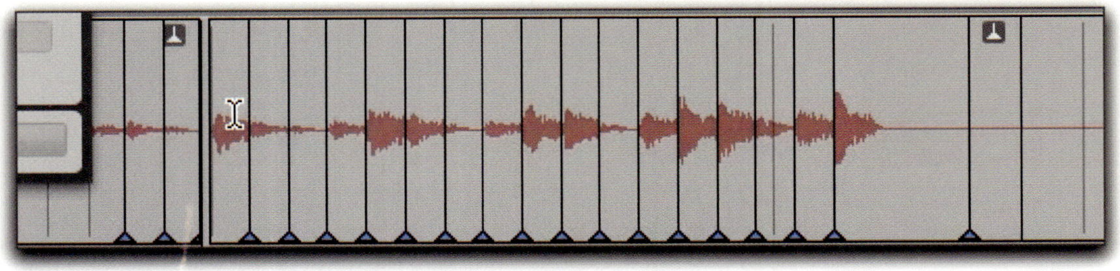

Figure 9.24 Open the Event Operations Window and Quantize the Warp Markers.

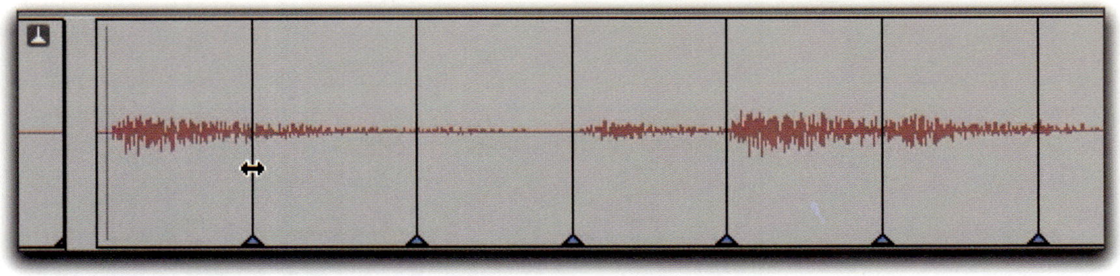

Figure 9.25 Fix Any Incorrect Warp Markers and Nudge the Transients into Place.

The more we pocket, the better our mix feels, and we're making a pretty radical difference at this point. That said, let's jump back and fix the break at bar 25. Switch to Warp view on the acoustic guitar and Shift+drag the transient into place at 25|2.

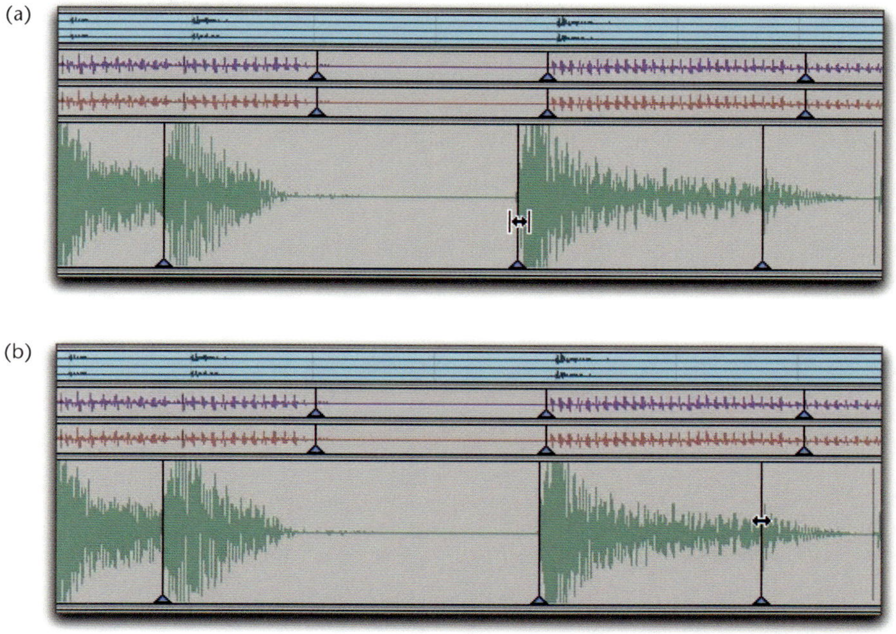

Figure 9.26 (a, b) Fixing the Acoustic Guitar Break.

While we're here, the chunky guitar needs its raking note moved into place as well. Select, separate, and nudge it into place against 25|2.

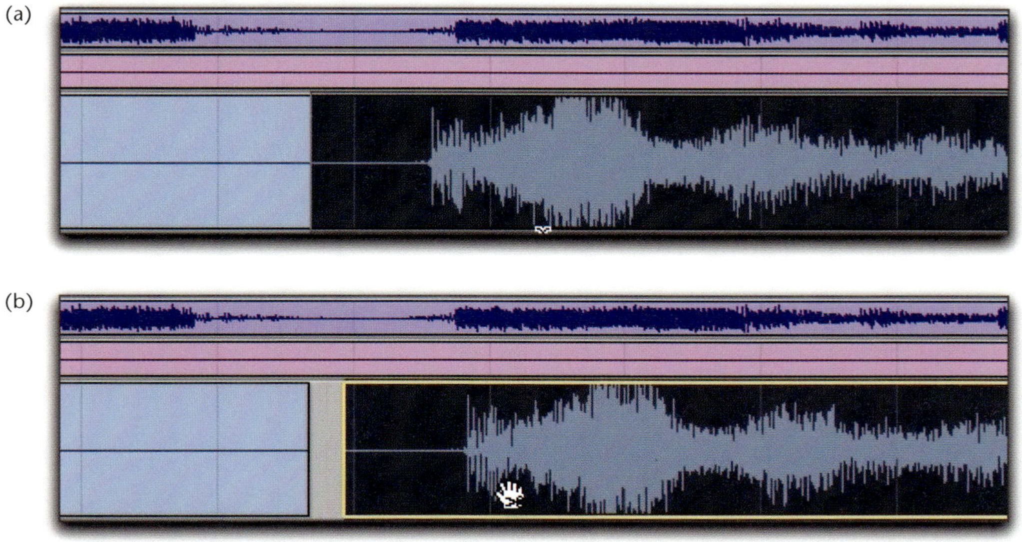

Figure 9.27 (a, b) Fixing the Timing on the Chunky Guitar Rake.

If you've done it correctly, you should have a much tighter performance that rocks faces. Of course, as we discussed earlier, a full-track pocket can take even an experienced editor upwards of 5 to 6 hours to complete to major label standards. If the budget provides for it, or you just want to beef up your skills, spend the time to make it great. You'll thank yourself.

In This Chapter

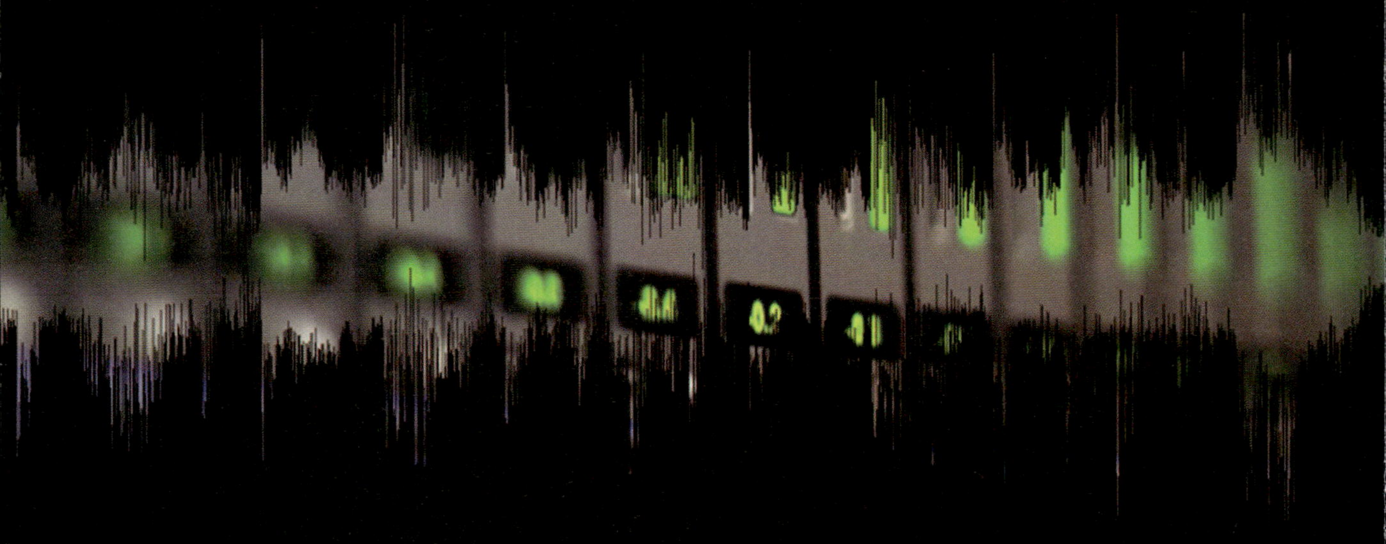

10

Premastering

Loudness Wars

Well, we've come a long way in equipping our mixing toolkit. For better or worse though, no modern mixer sends a track to their clients without at least a bit of premastering/loudness maximizing. While we, in the engineering world, lament the "loudness wars" that have been stamping the dynamic range out of music for decades, the fact is that if you give your clients a CD that sounds, significantly, quieter than a commercial CD, they'll tend to think that the mix sucks. It's just the world we live in, so let's make sure you know how to get the most volume out of your clients' reference mix, even if you send an uncompressed mix on to mastering.

Finally, it's time to remove the AIR Stereo Width plug-in that has been keeping us in mono-land for most of the mix. Ahhh … this is going to feel good.

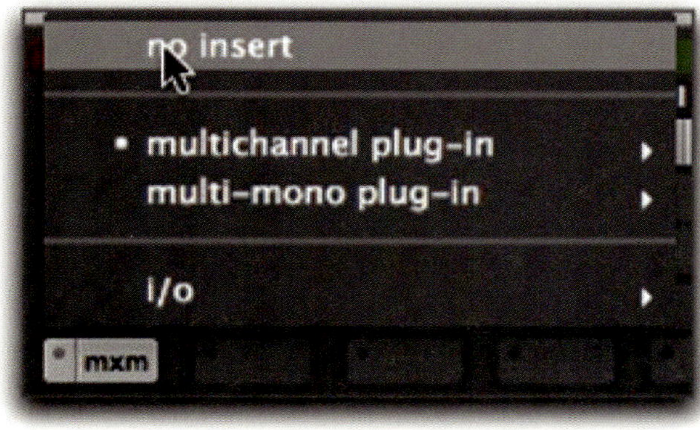

Figure 10.1 Be Gone Devil Mono Mixes!

© 2011 Nathan Adam. Published by Elsevier, Inc. All rights reserved.

This is your first time to really listen through with your timing, panning, levels and effects in place, in all their full stereo glory.

Figure 10.2 Bypassing the Mono-Izing Plug-in.

Job well done! If this were a book on every plug-in and piece of software I use in a daily session, I would have to mention amazing plug-ins like the SSL Bus compressor plug-in, as well as the infamous Izotope Ozone. Using these two plug-ins, you can get hot, screaming mixes with a minimum of fuss.

Premastering EQ

However, we can realistically get a good portion of that quality (albeit with a bit more effort) by using a chain of plug-ins that *is* included in Pro Tools 9. So, as a final series of steps, let's start with a seven-band EQ on the Master fader, and we'll follow it up with the Maxim limiter, and possibly a compressor.

As we always do, let's apply some corrective EQ first to clean up the ultra lows, and then add a touch of top and bottom end. Be sure to reference against similar commercial tracks in your CD library or even from the iTunes store if need be. As we do, I'm going to add a touch of Hi-frequency shelf (HF) and a touch of low-end bump with the LF EQ to the low end.

Using the same approach we always have, push the levels up higher than you need, and pull them back down to more modest gain increases of 1–3 dB.

Finally, engage a steep (24 dB/octave) HPF and roll the frequency up just high enough to cut out the lowest of the low junk frequencies. Mine got up to about 29.5 Hz. Again, reference the included Pro Tools sessions or the image on the next page.

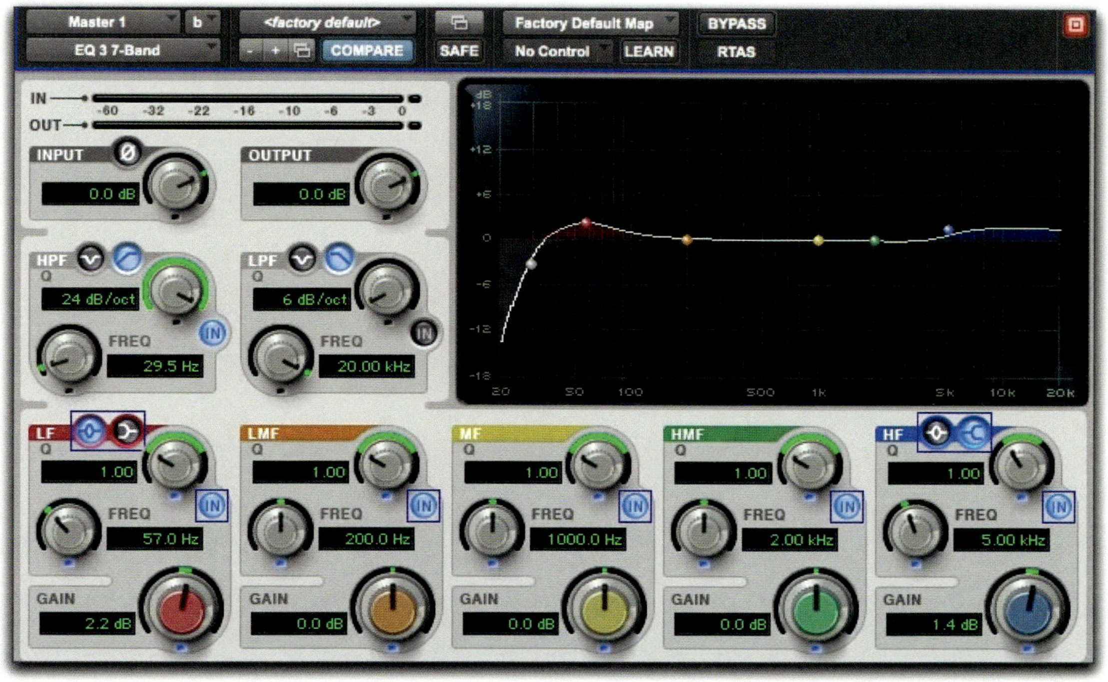

Figure 10.3 Our Final Master Fader EQ.

Pre- versus Post-Fader Metering

Before we go any farther, and just to avoid any e-mails from the nitpickers, you may be seeing the occasional red clip light firing in your session.

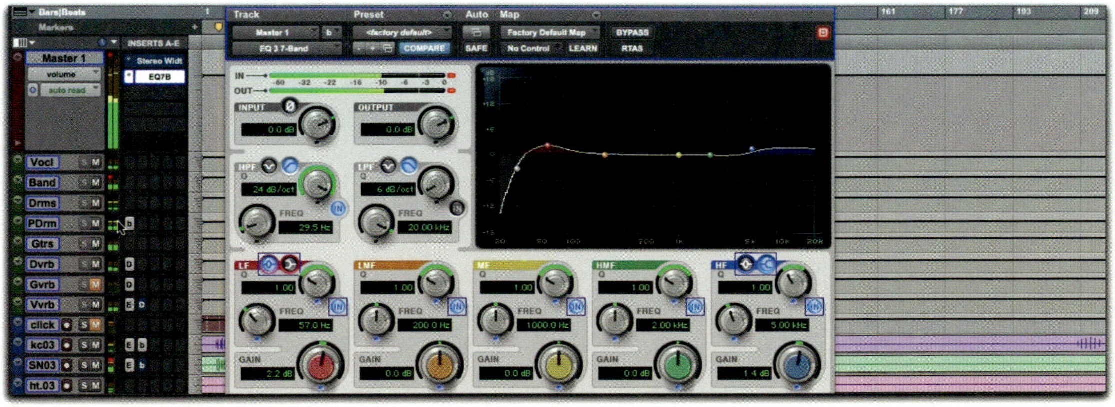

Figure 10.4 Red Lights Are Not a Mixes' Best Friend.

Before you go assuming you're definitely clipping your outputs, it may in fact be that you have your meters set to Pre-Fader Metering. Obviously, this shows the levels of your tracks before they've gone through any of the processing, plug-ins, or faders. So let's turn Pre-Fader Metering off (Options menu>Pre-Fader Metering) for a minute, then take one last look at the session to make sure there's no danger of clipping.

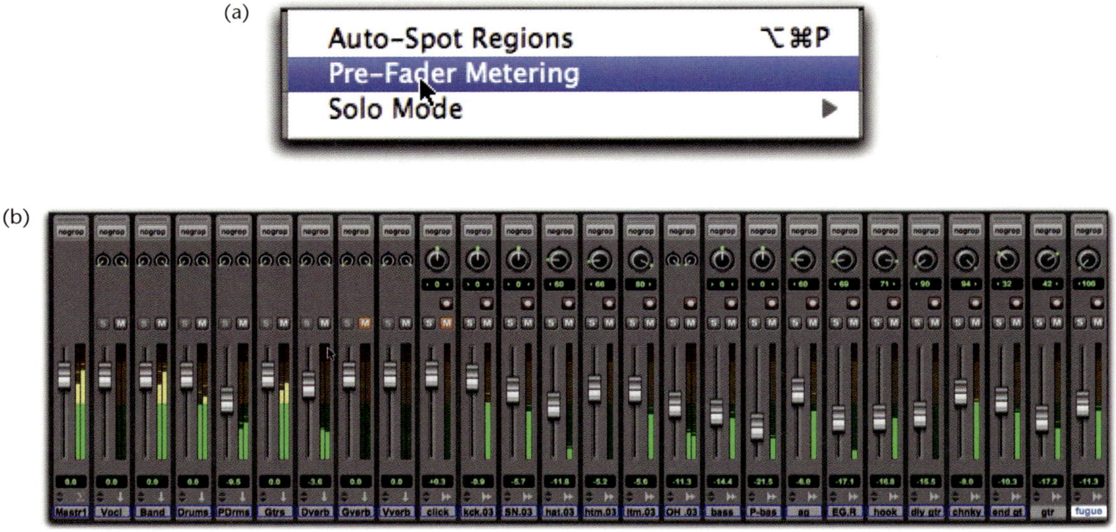

Figure 10.5 (a, b) Ahhh! No More False Clip Lights.

Maxim-izing Your Mix

Now that we've got everything approaching completion, we're going to put a Maxim brick wall limiter plug-in in the last slot of the Master fader. Because this mix will be primarily for our clients, we'll set the output ceiling as high as –0.2 dB so that they get a hot "louder must be better" mix.

Then, let's play back the track and start pulling the threshold down, until we hear just how dramatically this plug-in can blow up a track.

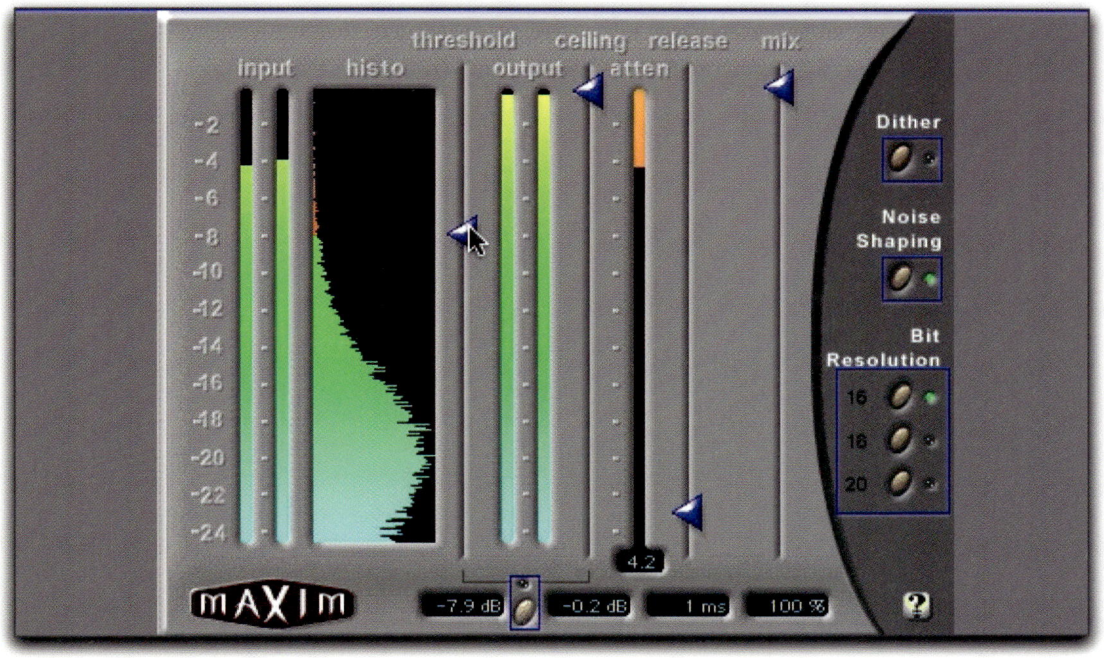

Figure 10.6 Laying the Smack Down with Maxim.

Even at –8 dB, we can both hear and see some pretty dramatic limiting starting to happen. The meter is currently showing 4.2 dB of gain reduction, with an audible increase in volume.

Identifying Overcompression

As a general rule of thumb, to avoid overcompressing the final mix (or the two-mix as it's called here in Nashville), gently pull down your threshold until you're starting to see the gain reduction meter, primarily, hitting on the mega transients like the Kick and snare drum. If you're seeing a nonstop gain reduction meter of greater than a couple of decibels, in all likelihood, you are squashing all of the dynamic range out of your mix. My threshold ends up around –7 dB.

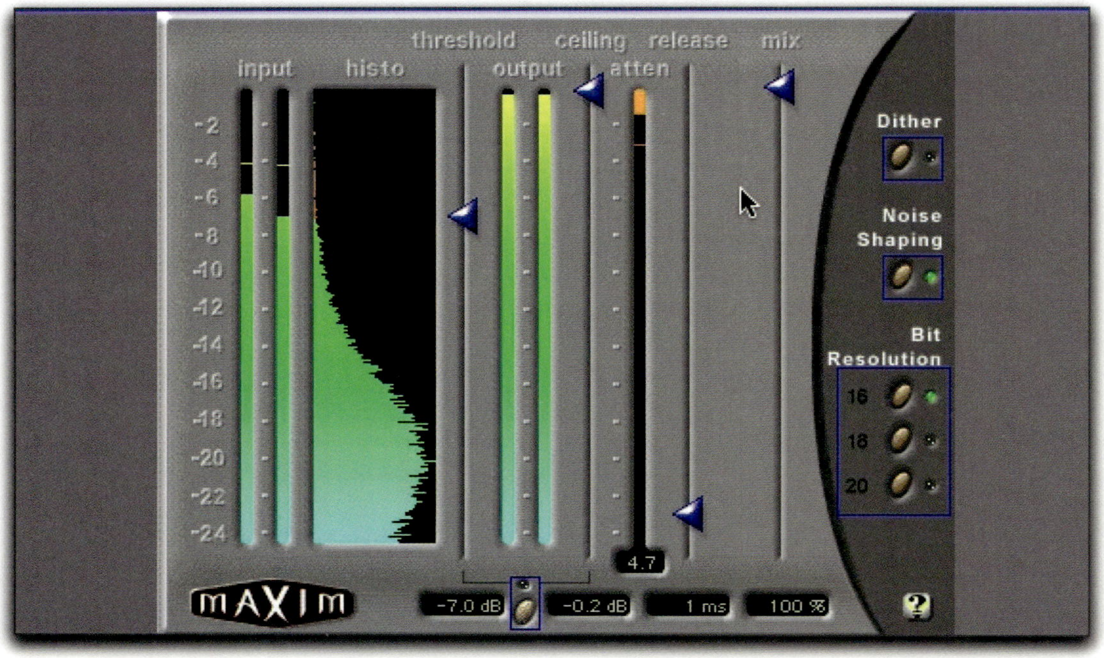

Figure 10.7 Final Limiter Settings.

What's really amazing is that because we've added a touch of compression and automation, repeatedly, to track after track after track, that over time with just a touch on a final limiter, we're already hitting radio ready levels. If you feel like the compressed sound is a bit overdone, you can always blend the Mix parameter in Maxim to less than 100%.

In This Chapter

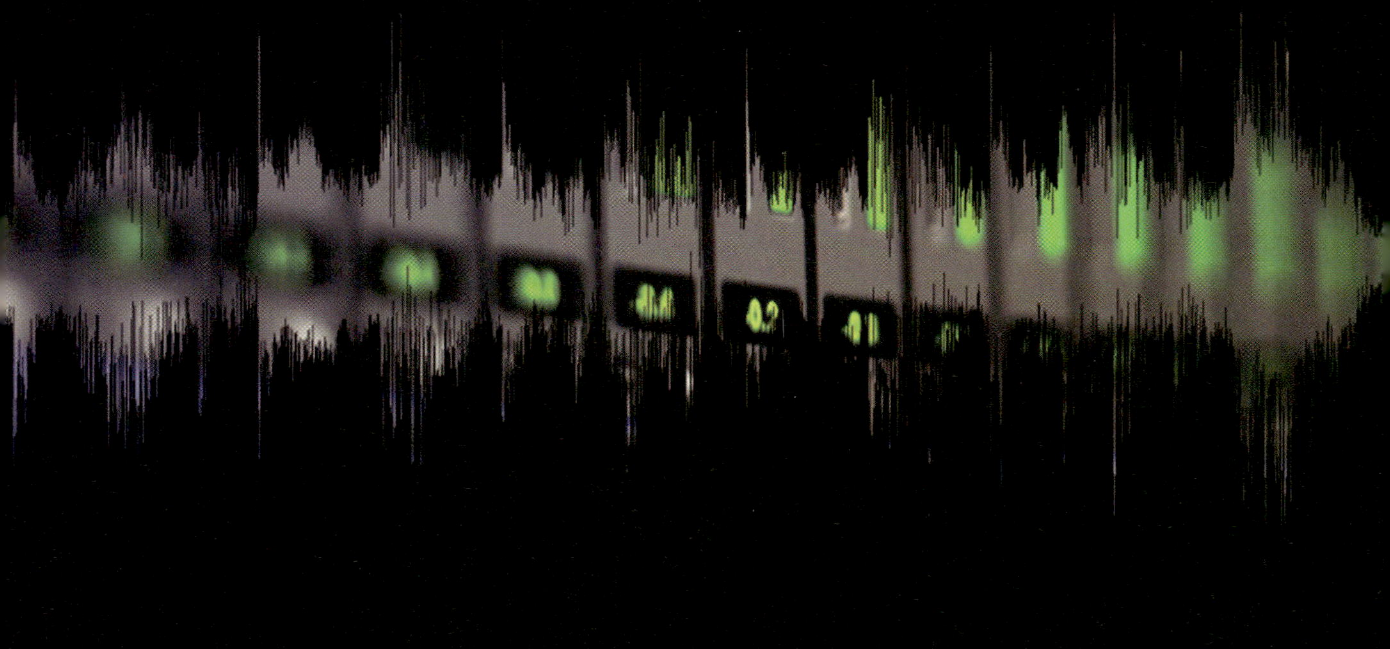

11 Final Mix

Master Fade

So now it's time for our piece de resistance, the master fade. Put your master fader in Touch automation mode and start playback around bar 129. Roll forward to the end of the song and pull the Master fader down in a logarithmic fade that lets the delays breathe out for a full repetition, but before any extraneous noise creeps in.

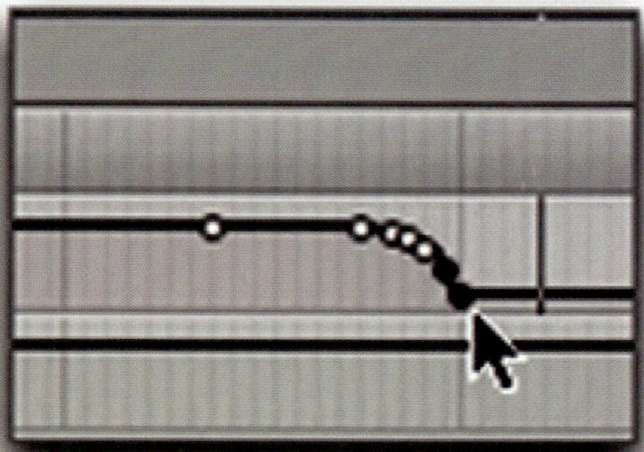

Figure 11.1 The Master Fade Out.

And there you have it. Take a final listen through your mix to hear just how far you've brought this garage recorded rock song. You just built it up over hundreds of tiny, independently miniscule decisions that accumulated to create a really rocking track.

So now it's time to bounce out our mix, ready to burn to a CD, upload to the web, or drop onto your phone.

© 2011 Nathan Adam. Published by Elsevier, Inc. All rights reserved.

Bouncing to Disk

My approach is to usually start with a marker point about 2 s before the intro of the song, and an out point just after the end of my master fade.

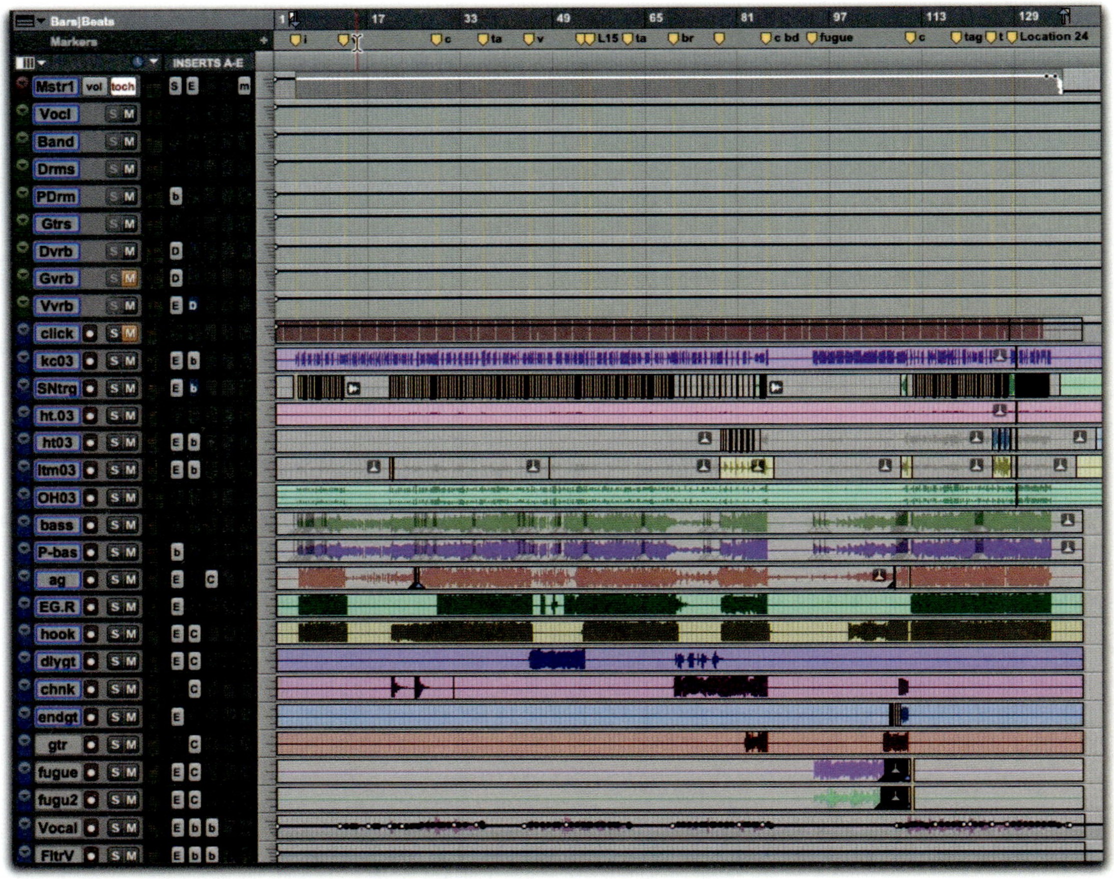

Figure 11.2 Selecting the Region to Mix Down.

Now selecting File menu>Bounce To>Disk (Mac: Alt+Cmd+B, Win: Alt+Ctrl+B) will bring up your Bounce parameters window. We've been mixing with output A 1–2 as our master output, and we want to use the industry standard Stereo, 24 bit/44.1 kHz Wav file for our master. You can leave the Conversion option on Convert after Bounce and click Bounce.

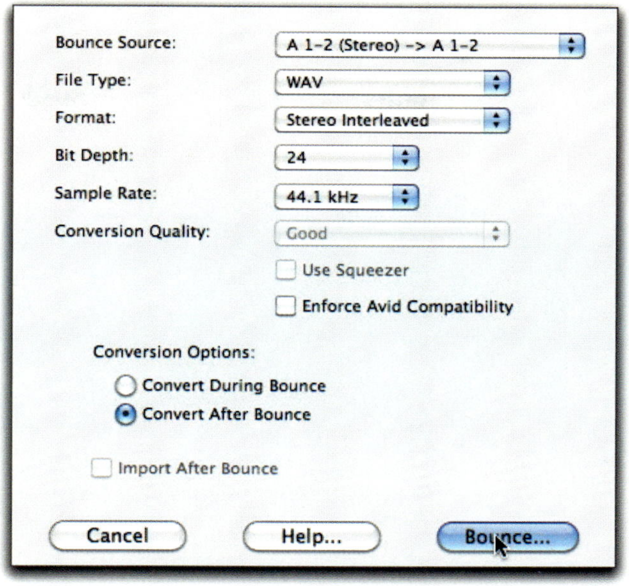

Figure 11.3 Final Mixdown Settings.

When Pro Tools asks us where to put the resulting mix file, I like to always make a new "Mixes" folder in my root session directory, as the first mix is almost never the final one.

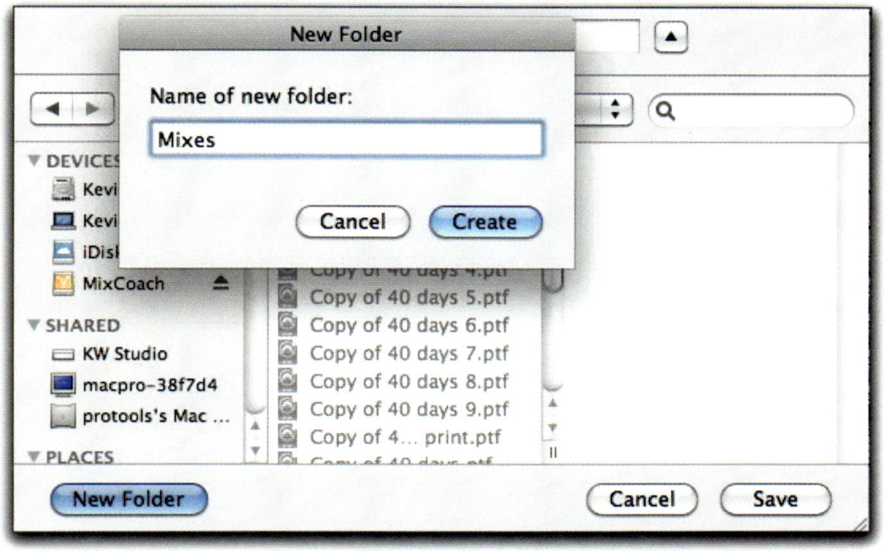

Figure 11.4 Making a "Mixes" Folder in the Session Directory.

Let's name it "40 days - mix" in accordance with NARAS naming guidelines and click Save to let the bounce begin.

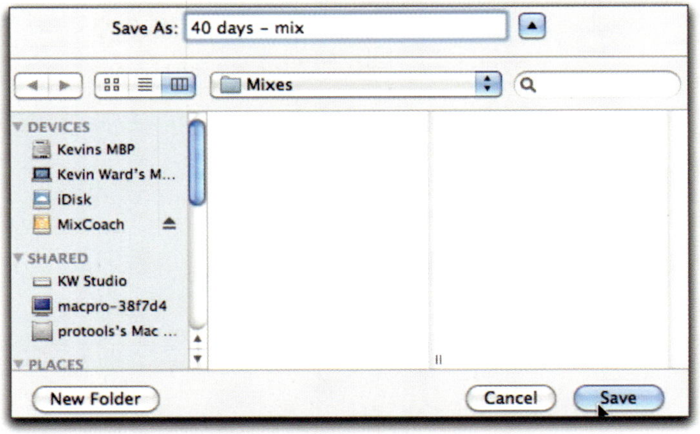

Figure 11.5 We've Worked So Hard to Get to This Point.

Closing Thoughts

And there you have it. Over the last couple hundred pages we have walked through how to use the mixing tools available in Pro Tools 9 to pocket, produce, and mix our way from the inside of a garage, to a rocking radio ready mix.

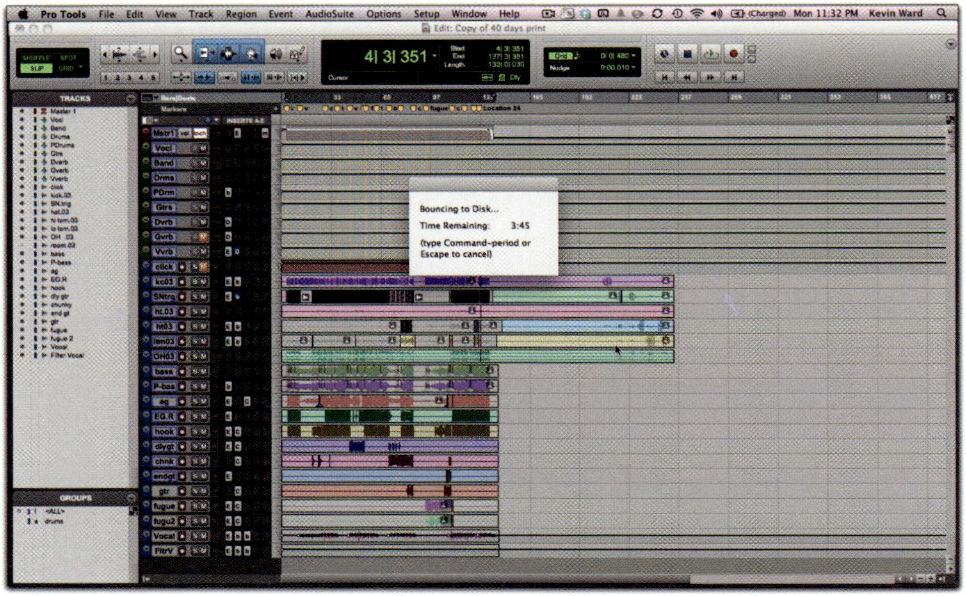

Figure 11.6 The Final Mixdown.

Without the assistance of A-level stunt players, a million dollar console, or a massive recording budget, you have filled your mixing toolkit with enough techniques and tricks to make any artist proud, and keep yourself employed.

We'd love to hear the mix you came up with as you developed your own techniques, so come visit us at www.mixcoach.com and pick up the final mix sessions, as well as more great mixing tips and tricks.

Happy mixing.

Nathan Adam and Kevin Ward

www.mixcoach.com

In This Chapter

12

Pro Tools Systems

Robby Resnick is a sales engineer for Sweetwater Sound and a longtime friend. From the moment we met, I had never met someone as into the minutia of gear as Robby. Because he's become an accomplished expert on setting up and configuring pro-level audio systems for everyone from home and project studios, to large-scale postproduction systems, I asked him to give us a clear guide on the new system and hardware options Avid gave us with Pro Tools 9. You can call him up directly at 800.222.4700. I do.

Finding the Right Pro Tools System for You

Choosing the right Pro Tools systems for you has never been easy. Previous to Pro Tools 9, the paradigm was simple; you could choose between Pro Tools LE and Pro Tools HD. Moreover, the lines between these two systems were easily defined. Pro Tools LE was the "limited edition." Digidesign (now Avid) placed software and hardware limitations on it. The number of tracks was limited, the import and automation modes were limited, and the processing was limited to that of your computer. It was either a 001 or a TDM system … and Digi life was simple. Then things began to change.

I won't bore you with the history of the MBox line or the 00X … and the M-Powered/M-Audio acquisition has recently become passé as well. I only bring them up because it's important to keep in mind that before November 2010, most would-be Pro Tools consumers were conflicted about what systems to get. Many people had questions like "Do I need HD?" or "What expansion toolkits would be right for me?" Then at the end of 2010, Avid announced Pro Tools 9 and the different configurations within. Suddenly, things became so simple, yet so difficult all over again.

At current standing, there are three basic systems to choose from Pro Tools 9, Pro Tools Native, and Pro Tools HD. Let's go over the details, pros, and cons of each system in hopes that it will help you select the system that is correct for you.

© 2011 Nathan Adam. Published by Elsevier, Inc. All rights reserved.

Pro Tools 9 Universal Needs

All Pro Tools 9 systems require a few basic things to be in place. I would recommend keeping them in mind as expenses while considering a system.

Computer

What would computer-based recording be without them? The machine you choose should be considered based on the Pro Tools Compatibility pages available on www.Avid.com. You computer will need to be on 10.6 (Mac) or Windows 7 (PC) and have access to the Internet. With that, you'll need a minimum one free USB port and an optical drive. Two hard drives that run at 7200 RPM or more are encouraged and a minimum of 4 GB of RAM is also strongly encouraged. Because these systems are reliant on a computer, the more capable the machine you have, the better off you'll be.

If you're unsure of what to get, please consult with a trusted recording professional working in the field. Many people fall into the pitfall of only speaking with a computer technician or computer sales person at a local electronics store and end up with the wrong machine because computer recording has special needs that most other people don't have to consider. If you're unsure, I'd recommend either purchasing an Intel-based Apple computer or contacting www.Sweetwater.com and purchasing a Creation Station PC (www.Sweetwater.com).

iLok

All Pro Tools systems require an iLok. These typically cost $60 or less, and some Pro Tools systems include it with the system. Please make sure you know whether your prospective system will have one already or if you'll need to purchase one separately.

Pro Tools 9 (Basic System): $699+

Pro Tools 9 is Avid's flagship software. It is a completely native software solution for recording. Similar to their competition, Avid has made this version of Pro Tools completely open for use with other manufacture's hardware. Any manufacture that develops a ASIO, WMD, or Core Audio driver can be accessed by Pro Tools 9. That said, Avid's proprietary link to their hardware (DAE—Digidesign Audio Engine) still remains. The advantages of DAE and using Avid hardware (like the 003 or MBox line) includes slightly lower latency, no dropping of samples, and recognition of the I/O within the software for easier routing.

Pro Tools 9 systems are highly capable. Out of the box, they will allow for 96 tracks, 64 instrument tracks, 512 MIDI tracks, 160 auxes, 256 busses, and up to 24 bit/96 kHz recording. I/O for this system is limited to 32 channels simultaneously. It requires a USB, FireWire, PCI, or PCI-e based interface to get this kind of I/O count. That said, the system will run on any computer that the iLok is plugged into, and a valid audio device is available.

These systems are 100% reliant on the computer that is hosting them. All OS overhead and driver overhead will affect the performance of your Pro Tools 9 system. These basic systems also have the highest latency of all other Pro Tools systems. Also, although the software is still somewhat limited compared with HD, these limitations do not effect most users. They include limits to the amount of video tracks available, limits in automation, and limitations on import and export options. With that said, to overcome these issues on a 100% native-based system, the owner may choose to add the Complete Production Toolkit2 ($1995). This will open up the software and remove the software limitations, enabling the software to run as the full version of Pro Tools HD software.

Pros: Low cost of entry, highly powerful, and economical when considering similar platforms, universal compatibility with other Pro Tools systems
Cons: Higher latency, limited I/O count, somewhat limited features, lower sample rates, inability to run TDM-based features
Who's it right for?: Singer\songwriters, voice-over talent, starting producers, composers, smaller home studios, portable systems, professionals needing a home\second system
Available options: Complete Production Toolkit, RTAS Plug-ins, Control surfaces such as the Command 8 and Euphonix MC series

Pro Tools HD Native: $5695+

In Q3 of 2010, Avid introduced its newest Pro Tools system: Pro Tools HD Native. This version of Pro Tools sought to bridge the gap between their existing solutions while leveraging the current advances in computer processing power. Basic Pro Tools systems have several limitations that this system seeks to solve. First, I/O count is limited to 32 (18 I/O previously). Also, the latency through a FireWire or USB is significantly higher than through a PCIe bus. Finally, the ability to use higher quality convertors in your chain has been limited by the digital I/O provided on "manufacturer X's" box. HD Native takes care of all of this.

HD Native is capable of 64 simultaneous I/O. This is accomplished by two 32-channel Digi Ports on the back of the card. Along with these ports, there is also a Machine Control port for slaving another recorder (typically an equipped tape machine). A Pro Tools HD Native system boasts significantly better specs than a standard Pro Tools system. It is capable of doing 192 mono tracks at 44.1\48 k. It can record all 64 I/O simultaneously and has room for 128 instrument tracks, 512 MIDI tracks, 160 auxes, 256 busses, and 64 video tracks. It requires the HD Native card, a qualifying I/O, and an iLok.

Unlike the smaller systems, Pro Tools HD Native does have HD exclusive functions. As mentioned earlier, normal Pro Tools systems can add a software pack to add HD software features, but not the hardware features. HD systems boast having track-specific input monitoring options and work with all Avid approved controllers all the way up through the Icon line.

As previously mentioned, Pro Tools HD Native relies on convertors that use the Digi Link ports. All of Avid's higher end I/O carry these ports; however, other manufactures such as Lynx and Apogee make hardware that is also capable of being the front end of these systems. Doing this has several advantages and disadvantages and it's best to talk to someone that specializes in building these types of systems to discuss the specifics. That said, it's important to understand that this port limits you to pro-level convertors. This will, in turn, raise the price of entry to this system, since many pro-level I/Os do not have a monitor section or preamps built on (aside from the HD OMNI, I can't think of one that does) … thus, you will need additional hardware to get it working.

This system boasts two very important features due to the HD Native Card. First, the latency spec is very low (~1.6 ms depending on convertor for a single mix utilizing the DSP on the Native card and assuming no plug-ins are present to cause additional latency). This is critical, especially in tracking applications. Secondly, the input monitoring option (auto input versus input only) is selectable on a track-by-track basis in HD. This is great for overdubbing and working with vocalists on overdubs and punches. It's a valuable feature that professional engineers have cherished in Pro Tools for years and is not something to be discounted as a deciding factor.

An added benefit for anyone involved in live sound and using a Avid VENUE console, HD Native is the first system in the line able to connect to VENUE for multitrack recording and virtual sound check features.

Like standard Pro Tools systems, these are entirely native based. They rely on a 32-bit floating point mix bus. Only RTAS plug-ins are compatible with HD Native. The price of just the HD Native card is $3495. Yet, in that card, consider this: Pro Tools 9 is $599 and Complete Production is $1995 … both of which are "included" in this pack. When using the Pro Tools license on the non-HD Native equipped machine, it shows itself as the HD software. Along with that, a pro convertor will run you at least $2200 (the most affordable on that, which will work, is Lynx Aurora 8 at $2200 with HD card).

Pros: Professional workflow, great latency specs, additional I/O count, works with professional convertors, low latency monitor mix, HD exclusive features
Cons: Higher cost of entry, limitations on I/O, inability to run TDM features and plugs
Who is it right for?: Budding professionals, tracking rooms
Available options: RTAS Plug-ins, additional I/O, control surfaces up to ICON Systems

Pro Tools HD: $10,000+

The Pro Tools HD system has defined our ideal workflow for digital recording. Even as a high school student and aspiring engineer, I knew that professional audio engineers were making entire albums within a computer. It fascinated me that they could track vocals with no latency, add effects to live sources, and edit hundreds of tracks at a time. The funny thing was, no one ever told me that it is not commonplace. So I'm here to tell you, that the Pro Tools HD system *is* the system you envision when you think about digital recording power.

While not a new concept, Pro Tools HD is the latest incarnation of a legacy of DSP-based recording solutions whose lineage began with the late and great Sound Tools on the Apple II using NuBus cards. Needless to say, they have had a while to perfect this.

Pro Tools HD is a fully scalable professional recording solution. It is based on proprietary processor cards (referred to as "Core" and "Accel" cards). Each card consists of several Motorola processors, dedicated memory, and a PCI or PCIe extension for interfacing with the computer. The smallest of these systems is a Pro Tools HD 1 or Pro Tools HD Core systems. Additional Accel cards are added to provide additional processing (for track count, sample rate, I/O, or plug-in count). These Accel cards link to the Core via a ribbon cable. It is important to note about these systems that the entire audio engine and mix bus live on these cards, relieving the computer of this duty. A major advantage to this, aside from added CPU power, is processing depth. Unlike the above systems, Pro Tools HD processes in a 48-bit fixed-point mix bus, which sonically is nearly equivalent to 64-bit floating point. This enables HD systems to have better dynamic range and clarity than standard Pro Tools systems. Additional Accel cards can be added, allowing for more tracks and effects. These cards always communicate directly with the Core card. They only draw power through their PCI or PCIe connection—not passing data. The maximum supported system is an HD7 (Core+6 Accel), though I've heard of larger systems in the field using specialized configurations.

Pro Tools HD boasts the most impressive specs of all Pro Tools systems. It shares many numbers with its Native cousin, including track count, MIDI Tracks, Instrument Tracks, and so on. The differences are:

I/O count: Each card accommodates 32 channels of I/O. That said, Pro Tools is "limited" to 160 channels of active I/O.
Latency: ~.07 ms depending on convertors.

The most significant feature of Pro Tools HD is the ability to use TDM plug-ins and features. TDM plug-ins are plug-ins coded to specifically run on the cards themselves. Due to the nature of Alpha Processing, these plug-ins appear to happen in nearly real time. TDM plug-ins are often used during the tracking phase as well as the mixing phase. These plug-ins have, in some cases, taken over the need for external processors and are available from Avid as well as a wide variety of third-party companies. There are a couple of other features that are TDM reliant. All delay compensation on an HD system is handled on the cards. Even one of Avid's most recent feature releases was TDM based. HEAT, which is Avid's analog console emulation feature in Pro Tools 9 is TDM only. For those like me, who have heard and worked with HEAT, the difference is quite stunning. For some people, this ability may be a make or break in itself.

And HD1 system will run $8000 for the card itself and Pro Tools 9. Along with this, a professional I/O is needed to make the system functional. Personally, I rarely recommend to a client a Pro Tools HD 1 system. In my eyes, a single card is easily monopolized by general workflow considering a fair track count (32 voices) and a handful of plugs with delay comp. Thus, I often recommend an HD2 system to start with for real HD. Still, it's best to consult a professional before taking this type of jump into a system to make sure you get exactly what you were looking for.

Pros: Fully scalable, highly powerful, nearly zero latency, no locked features, higher definition processing for audio, lighter processing load on host computer, HEAT
Cons: High cost of entry, additional cost of TDM plug-ins, requires modification of computer internally
Who is it right for?: Anyone looking for the best experience in digital audio. From home studios to professionals, any engineer can appreciate the seemingly limitless capability of Pro Tools HD systems
Available options: RTAS and TDM plug-ins, HEAT, additional I/O, control surfaces up to ICON Systems

As previously mentioned, it is always wise to consult with a professional before purchase. There are a lot of details to take into consideration.

Finally, a question many people ask me is "What computer should I buy to run Pro Tools?" I always advise the most powerful within your budget, which is qualified to work properly. The reason being that, all of these systems ultimately rely on the power of the host computer. You wouldn't want to drive an in-line -4 203 HP Corvette, would you?

When assembled properly, any Pro Tools system can be a powerful tool for music creation, sound design, and audio restoration. Assembling a system is intended to be a fun and educational experience. Make sure to do your homework and find someone you can trust for proper guidance. Above all, never forget that these are simply tools for creation. Like a carpenter, make sure you understand how all your tools work and which tools are appropriate for each task. Never let the tool stand in the way of getting the job done correctly.

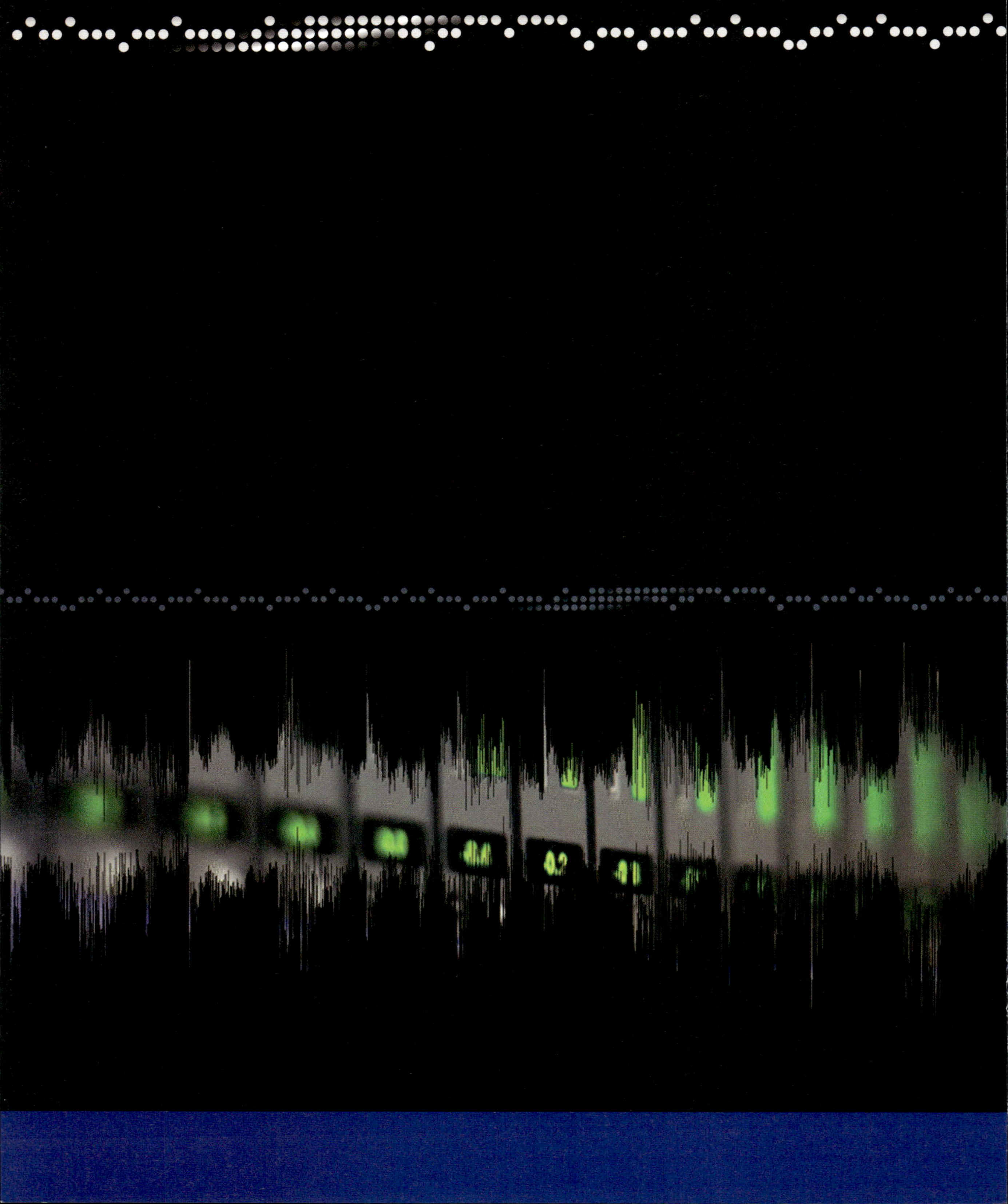

Index

Page numbers followed by *f* indicates a figure and *t* indicates a table.

W

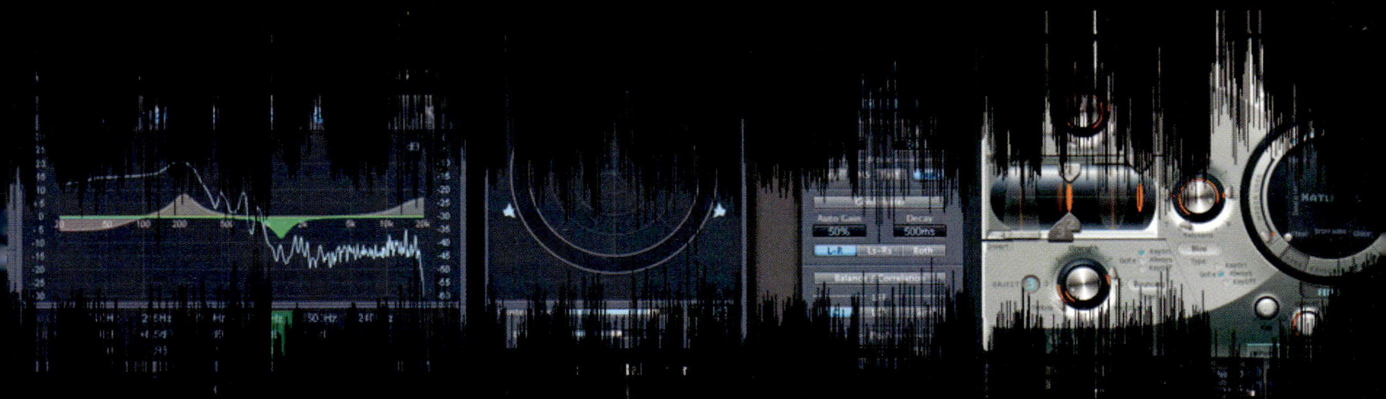

Bonus Chapter

From *Logic Pro 9: Audio and Music Production*

(ISBN 978-0-240-52193-0)

From initial track laying through to mixing, sound design, and mastering, Mark Cousins and Russ Hepworth-Sawyer bring you *Logic Pro 9*. By highlighting the relevant parts of each application, they take you through every step of the music creation and production process, giving you all the tips, tutorials, and tricks that pros use to create perfect recordings.

The book has full-color screenshots illustrating the tools, functions, and the new look of Logic Pro 9, and the companion website has audio samples and loops.

Logic Pro 9 covers more than just the software. It will help you make the most out of every recording session, and will illuminate and inspire your creative and sonic endeavors!

© 2010 Mark Cousins and Russ Hepworth-Sawyer. Published by Elsevier LTD. All rights reserved.

In This Chapter

Plug-In Boxouts

Plug-In Focus

Knowledgebases

Walkthrough

Logic Tips

8

Mixing in Logic

8.1 Introduction

Up until a few years ago, there was only one accepted way of creating a professional sounding mix: hiring a commercial studio facility and mixing through a traditional mixing console, like an SSL or a Neve. However, times have changed significantly since then, and mixing "in the box" – in other words, mixing completely within the domain of your audio sequencer – has become a viable, and some would say more flexible, alternative to the traditional console route. In truth, applications such as Logic offer a tremendous amount of creative and technical freedom to today's mix engineer: complete and instantaneous recall, compression and equalization on every channel, studio-grade reverb, full automation, and much more besides.

In this chapter, we're going to take a closer look at the process of mixing in Logic, looking at both the technology it has to offer to create an effective mix and how we can knit these elements together as part of the mixing workflow. This is also an opportunity to get to know many of Logic's essential signal processing plug-ins in a more informed way, as well as taking a more detailed look at how the Mixer area works – aspects that can actually improve the entirety of your workflow in Logic, as much as the mix. Looking more closely at the process of mixing itself, we'll explore how to use compression and equalization to create separation, as well a bus processing (on aux channels), and of course, spatial treatments like reverb and delay that help define the depth and dimension of your mix.

Beyond the essential signal processing tools of mixing, we'll also take a look at features like automation, grouping, and folders, all of which help make your mix easier to manage, especially with the large track counts that most projects seem to encompass.

© 2010 Mark Cousins and Russ Hepworth-Sawyer. Published by Elsevier LTD. All rights reserved.

8.2 Channel Strips: Understanding Your Virtual Console

Before diving headlong into your first mix on Logic, it's worth taking some time to distinguish between the various channel strips that the Mixer comprised and how these various elements interact with each other. Generally, a good mix in Logic makes use of all the features of the Mixer – channel inserts, bus sends to aux channel strips, output masters, and so on – so it's worth familiarizing yourself with their features in much the same way as an engineer would "get to know" the sections of a physical console.

Audio Channel Strip

An audio channel, as we've already seen in previous chapters, governs the basic path in and out of Logic for a recorded signal. In a mix, audio regions in the corresponding track will be sent down the channel strip, through the various insert processors (compression and EQ, for example) and bus sends (for reverb and other effects) to a designated output.

Instrument Channel Strip

The instrument channel strip duplicates the same features as an audio channel strip; only this time you get to work with the signals generated by the virtual

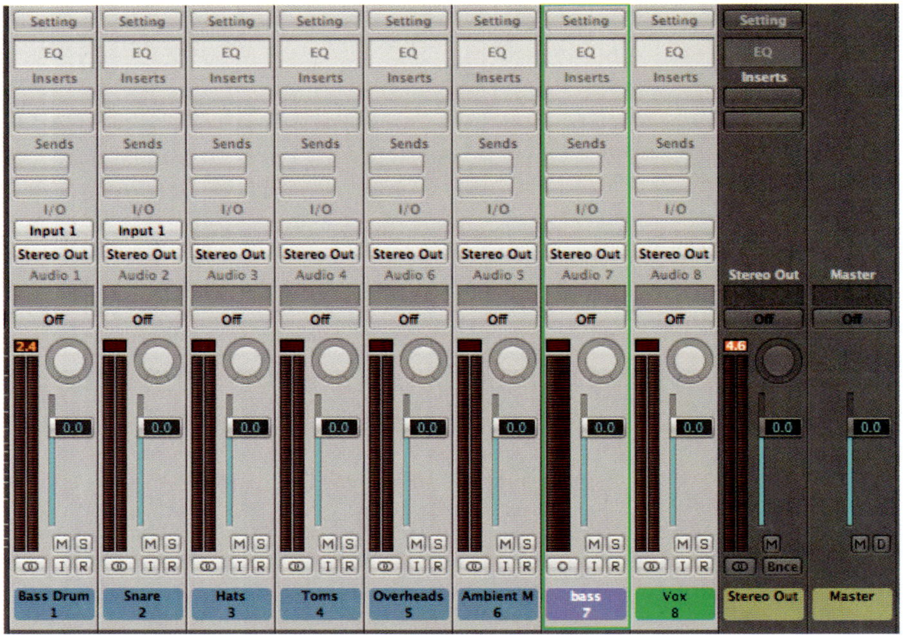

Figure 8.1 An audio channel strip.

Figure 8.2 An instrument channel strip.

instruments in your session. In addition to this simple application, an instrument channel strip can also be configured for Multi Channel operation so that multiple outputs from the same instrument (like the EXS24 or Ultrabeat) are sent to a number of additional auxiliary output channels. Besides having individual level control, this also allows you to apply different effects onto each output so that a snare, for example, might have different equalization, compression, and reverb to that of a kick drum.

Plug-In Boxout 1

Distortion Effects

Logic includes an impressive set of effects dedicated to low-fi transformations and crunching-up sounds – from the digital extremes of Bitcrusher, to the warmer tones possible with Guitar Amp Pro to the detailed emulation of many favored tones emulated within Amp Designer. Besides being an obvious addition to guitar parts, distortion can also be a great source of

(Continued)

color and interest on any number of sounds on a mix. Try using distortion to add some grit and body to a drum loop, for example, or a touch of drive on an aggressive vocal.

The first choice of amp-like distortion would have to be Amp Designer, which provides immense flexibility with respects to the choice of emulated head, speaker cabinet, microphone type, and its position. Additionally, there are a considerable wealth of presets to get you stared. Add to this the power offered by the new Pedalboard plug-in offering the guitarist endless scope.

Figures 8.3 and 8.4 Logic's Amp Designer offers an impressive selection of amplifiers to choose from.

Figure 8.5 Amp Designer with Pedalboard brings a whole new scope for guitarists in Logic 9.

Guitar Amp Pro, however, is still a solid option, which emulates the Amp, EQ, and speaker components of a typical guitarist's amplification system, as well as modeling a number of different mic'ing options. The clear advantage here, as with Amp Designer, is that the tone of the guitar can be completely modified (changing the Amp head, for example, or the position of the virtual microphones) at any point in the production process, without having to rerecord the guitar.

Distortion, overdrive, and clip distortion come from earlier incarnations of the application and tend to be less suitable for applications on electric guitar. They are, however, an excellent addition to drum loops or acidic synth lines generated from the ES2. Both distortion and overdrive have particularly simple controls with just a drive and tone parameter, alongside a corresponding output reduction slider to avoid ripping your monitors to shreds!

The prize for the most bizarre distortion effects goes to the Phase distortion and Bitcrusher plug-ins. Phase distortion sounds great on drums loops needing to be mashed up beyond all recognition. Based on a modulated-delay line, it can produce tones that range from a warm fuzz to the sound of a FM broadcast gone considerably wrong! Bitcrusher, on the other hand, is a great source of digital grit – especially as you reduce down the bit slider (to around 8 bits) and start applying some heavy downsampling (103 and beyond).

(Continued)

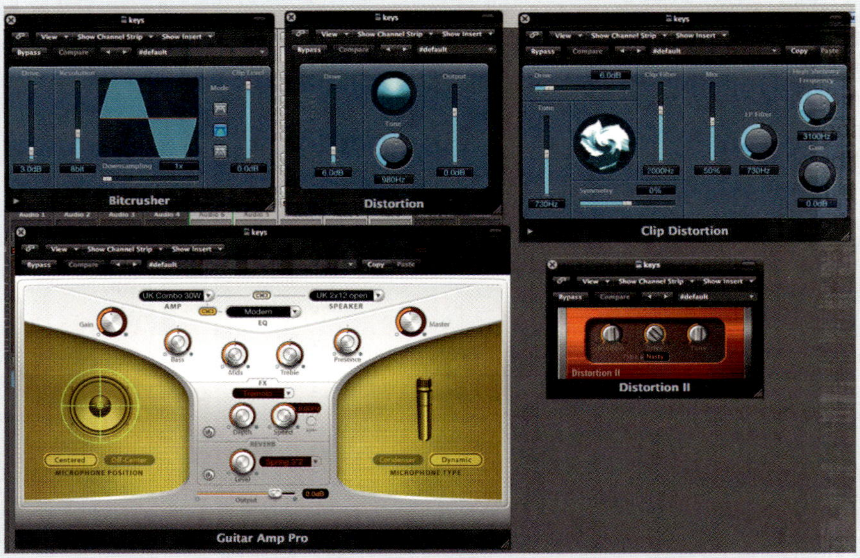

Figure 8.6 Distortion can bring plenty of color and grit to your mix. Try experimenting with Logic's wide range of distortion effects to achieve anything from a subtle valve-like warmth through to early digital distortion. Guitar Amp Pro still provides solid guitar tones in addition to Amp Designer. Logic's Amp Designer offers an impressive selection of amplifiers to choose from.

Aux Channel Strip

The humble aux channel strip is one of the most useful parts of Logic's Mixer and the source of a number of different techniques in mixing. First, it can be used as means of applying send effects like reverb or delay – anything where you want to actively control the balance of wet and dry sound using a separate fader. In this example, the aux channel strip has the effect strapped across its insert path with instrument channel strips or audio channel strips sending signals via designated bus sends. This allows any number of channels to access the same reverb, for example (an essential way of preserving DSP resources), as well as being able to control the return of the effect in the same way as any other audio signal in the mix – in other words, it could be equalized, compressed, and controlled by movements of the fader.

Another less immediate, but altogether just as useful, technique is to use an aux channel strip to combine a number of different channels (including audio and instrument channels) on a single fader. For example, many engineers using a traditional console will create a submix of the drums to a selected bus master fader

Figure 8.7 An aux channel strip.

(in this case, one of Logic's aux channel strips), allowing them to quickly control the level of the drums relative to other instruments in the mix. Additionally, they could also make use of inserts on the bus faders to apply compression, EQ, and so on to the entirety of the drums, rather than individual channels within it – something which can just as easily be done of Logic's aux channel strips.

The final important role of the aux channel strip is to operate as a means of inputting external signals into the mix. If your audio interface supports enough inputs, there's no reason why you can't patch-in external compressors, equalizers, synthesizers, and so on directly into your mix. Although you might need to account for a small amount of latency (as discussed in Chapter 3), using dedicated recording hardware can often supply much more character than conventional plug-ins. Of course, given the provision of inserts on these aux inputs, there's no reason why you can't also use Logic plug-ins on top of whatever device you're inputting into that particular aux input.

Stereo Output Channel Strip

The stereo output channel strip represents each of the physical outputs in your Logic system. If you're using a simple two-in, two-out USB audio interface, you'll only see one of these faders. However, if you're working with a multiple output FireWire or PCIe soundcard, you can increase the number of output faders corresponding to each physical output. On the whole, most mixes simply use one designated output (1–2) as the "destination" of the two-track mix. If you intend to render the mix directly from the Logic session, you'll also need to make use of Bounce button, found on the output channel strip or File > Bounce or simply Cmd + B (for more information on bouncing to disk, see Chapter 9).

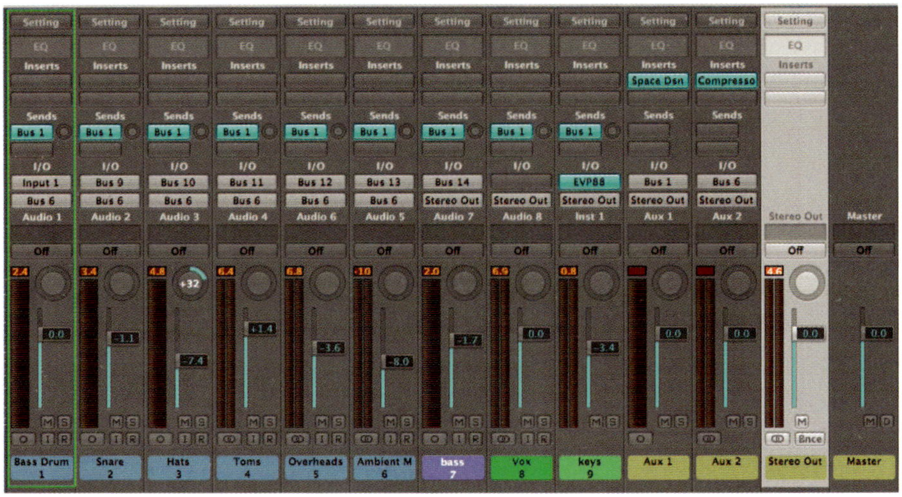

Figure 8.8 An output channel strip.

Plug-In Focus

To Gate or Not to Gate … That Is the Question

Music trends change, so it's not a surprise that the process of recording also adapts concurrently. For example, just a few years ago, it was commonplace to gate the components of a drum mix (kick, snare, toms, and so on) to within an inch of their existence! Nowadays, it seems that listeners, engineers, and musicians all seem to prefer a looser sounding kit, resulting in much less use of gating.

Now and again, though, you may well feel the need to improve your aural hygiene – maybe a vocal microphone has some obtrusive background noise made worse through the application of compression or that one and only guitar take has a little too much noise from the amp. If you're pushed for time, there's no doubt that the noise gate plug-in remains a quick and effective way to attenuate problematic noise in between notes – although remember to gate before the compressor and not after it! The trick with setting the noise gate is to start with hard settings (quick Attack, Hold and Release, with 100 dB of reduction) to find the right threshold. Although the gating sounds harsh, you'll be better able to find the "sweet spot" just above the amplitude of the noise. With this set, back off some of the settings, especially Release and Reduction to soften the effect – surprisingly, even a small amount of reduction (6–10 dB) can have a big effect on the overall cleanliness.

If you've got a little more time, most users now tend to approach the gating issue through a few crafty edits. Try using the Arrange window's Strip

Silence option (Audio > Strip Silence) as a means of Logic preparing the majority of edits for you. As with a conventional noise gate, you'll need to establish the right threshold to get the best results, although arguably this is easier to "see" on Strip Silence than it is to "hear" on the noise gate. However, the big advantage with the Strip Silence approach is that you can modify the edits later on, making sure none of the important attack transients are missed out.

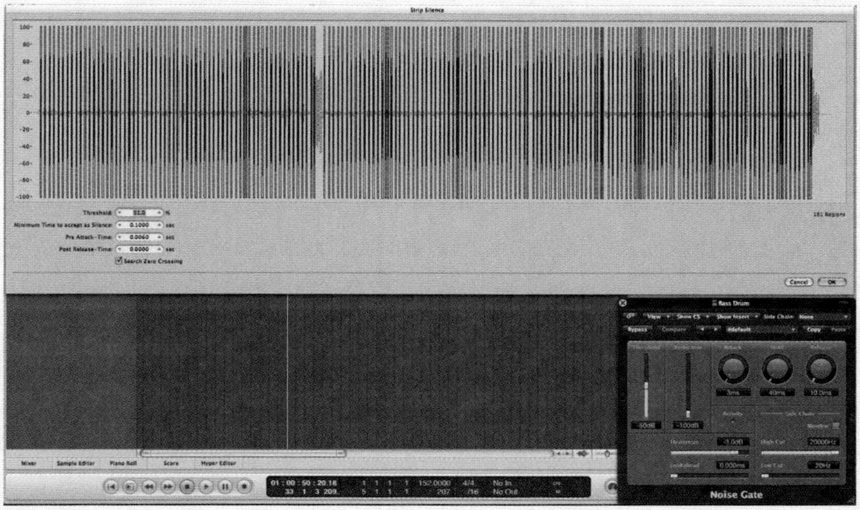

Figure 8.9 Two choices for improved aural hygiene – the traditional noise gate or the current favorite, Strip Silence.

MIDI Channel Strips

Arguably, MIDI channel strips should be kept distinct and separate from the audio mixing, but given their inclusion as part of the Mixer, it's worth clarifying their exact role. Unlike all the other mixer channels in Logic (which control the flow and qualities of audio within the application), the MIDI channel strips have no direct control of the "internal" properties of your mix, but instead they work as controllers for external MIDI hardware connected to your MIDI interface. However, in the case of the synths and sampler being returned via aux channel strip, these MIDI channel strips could have a direct effect on your audio mix – maybe balancing 16 MIDI channels being returned to one stereo input on your audio interface.

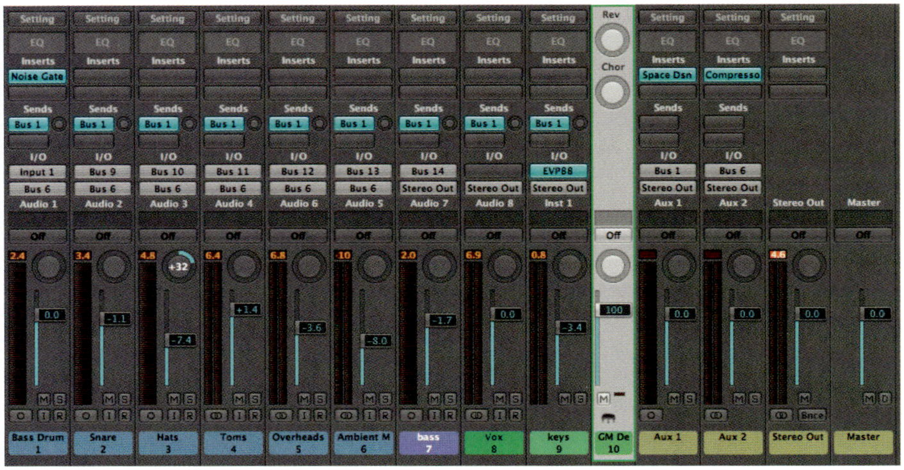

Figure 8.10 A MIDI channel strip.

Master Channel Strip

The master channel strip provides a quick-and-easy access point to the level of signal going to all main outputs and is directly linked to the "volume" control as part of the features on Logic's transport bar. In situations without control room monitor levels, for example, this could be used as a means of adjusting the monitoring level, although it should be noted that any digital bounces made in Logic will be subject to the master channel strip's gain adjustments.

Note that you cannot insert plug-ins across the master fader in stereo mixing (instead use the Stereo Out strip), with the master channel simply being used as a level attenuator. In surround mixing, however, its role – and the role of the outputs – slightly changes, which we'll cover in more detail in Chapter 10.

8.3 Organizing Your Mixer: What You Do and Don't See

Ask any good mix engineers about the fundamental tool that helps them negotiate an effective mix workflow, and they'll often reply "a well-organized mixer." On a traditional console, for example, engineers might decide to repatch the tape returns so that the order and arrangement of the various channels best reflects how they might want to carry out the mix – maybe with the drums in the first 8–10 channels, for example, while bass, guitars, and keyboards following on subsequent channels. So, before you start diving into equalization settings and designing your soundstage, take some time to organize both the arrangement and track list to help navigate your way around the mix successfully. Logic also provides a number of ways to organize and structure the appearance of your mix, for example, separating off sections, hiding unwanted channels, and viewing the complete signal path.

Mixer Views

The first concept to grasp with respect to using the Mixer interface effectively is the link between the Arrange window and the Mixer window. By default, the Mixer is set to its Arrange view out of the three possible view modes at the top of the Mixer window – Single, Arrange, and All. In Arrange mode, the order

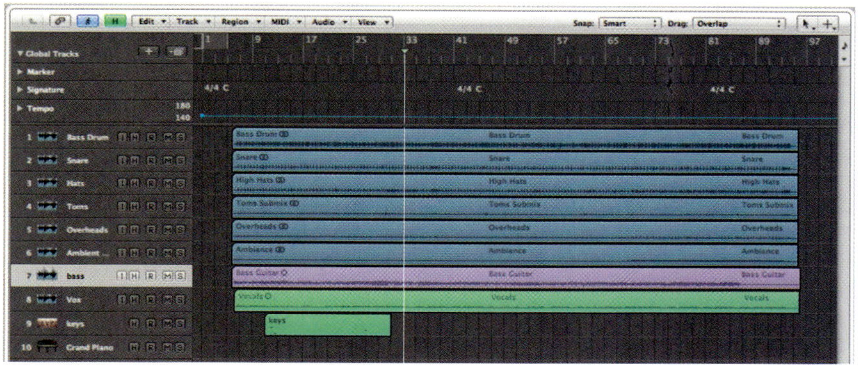

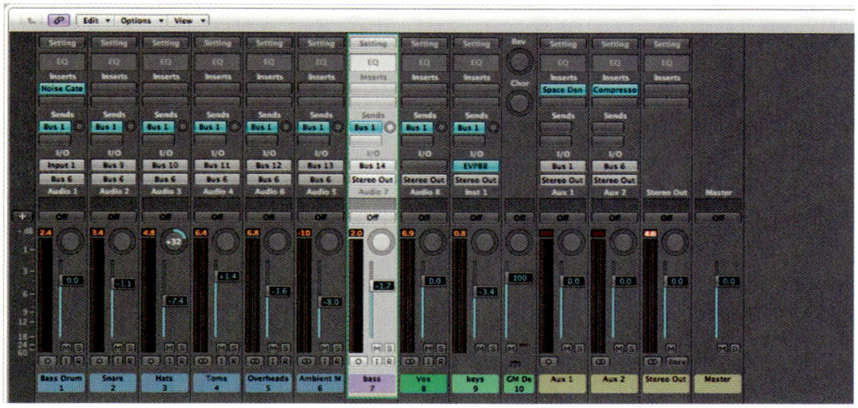

Figures 8.11 and 8.12 In Arrange view mode, the Mixer area directly corresponds to the order of the channels in the track list. The Hide Track option can also be a useful way of removing further channels.

and amount of channels displayed directly corresponds to the Arrange track list. For example, you might have reorganized the vocals to place them on the first track lane – a decision that also impacts on the arrangement of the Mixer. So, one simple way of organizing your Mixer is to organize your track list!

Following this concept, you'll notice that the track Hide feature that we first saw in the editing chapter also carries through into the Mixer. One intriguing difference to the behavior in Arrange window, though, is that the Hide option has an immediate effect on the currently viewed channels, irrespective of whether the H button in the top of the Arrange area is active or not.

An alternative to the Arrange view is the All mode, which displays the entirety of audio channels in your Logic project arranged sequentially by their assignment (Inst 1, Inst 2, and so on), rather than their order in the Arrange area's track list. This is a useful way of seeing the entirety of your mix, including objects (like the metronome and Prelisten channel) that you wouldn't usually have access to. However, without the ability to organize this view, it is easy to become disorientated, especially in bigger mixing sessions.

Knowledgebase 1

Clipping Faders: Good or Bad?

On the whole, conventional engineering wisdom tells us to avoid situations where a signal "clips" the meters – in other words, whenever we see one of Logic's channels faders or master outputs go into the red. However, in reality, there are some situations in Logic where this is acceptable and others where it demands your immediate attention!

In Logic, a fader registers a clip whenever it is presented with too much signal – this might be because you've brought up the output gain of the compressor too much, or that you're feeding too much signal to the main output channels. Looking first at channel faders and instrument tracks, the clipping we see here isn't particularly problematic, with no immediate chance of you hearing distortion at first. This is all because Logic's Mixer incorporates a degree of safety margin – or headroom – built in, using an enhanced bit-depth resolution to tolerate peaks in the region of 11–16 dB.

Moving onto the main output channel strip, however, things are not quite so flexible. Any final digital medium has a fixed amount of level it can tolerate – in other words, a CD or WAV file cannot exceed 0 dBFS (0 dB on the output fader) without the top of the waveform being clipped or distorted. In this respect, it is important to ensure that the main output fader doesn't ever go into the red when you print off your final mix (although the odd peak

Figure 8.13 Although clips on channel faders aren't too problematic, it is advisable to avoid clipping the main output.

during the mixing process isn't too disastrous). Given the amount of signals potentially feeding this channel, though, this is quite easy to encounter.

So, if your output channel does clip, what should you do to resolve this? The first and easiest solution might be to turn down the faders contributing to the mix, which could be done by placing them all into a group, and then turning the group down accordingly. If it's in the region of a few decibels, you could also choose to turn the master channel strip fader down; although if you're using it in the realms of 2–6 to 2–10 dB, then it suggests that there is a fundamental level mismatch that needs to be addressed. Another option is also to put an instance of the gainer plug-in (on the master output) and also use this to reduce to final mix level ahead of it being bounced.

Viewing by Signal Path

One of the most interesting view modes is the so-called Single mode, which shows the complete signal path of a selected channel. The signal path illustrates the full journey a channel makes through Logic's Mixer, including all the associated bus send effects, any submixing via aux channel faders, and of course, the final output channel strip. Using Single mode is a great way of honing in on the specific signal-processing characteristic of one sound in the mix, especially where it's using multiple bus sends (for reverb, and so on) that could be positioned well along the Mixer in the standard arrangement view.

Dropping Out Sections of Your Mixer

Using the various tabs along the top right-hand corner of the Mixer area, you can switch out various corresponding sections of your Mixer. For example, you might choose to remove MIDI faders given that they don't directly input on the signal processing aspects of the mix. Alternatively, by removing the audio and instrument (Inst) from the view, you can gain quick access to the aux masters and main output without having to scroll across the full length of the Mixer.

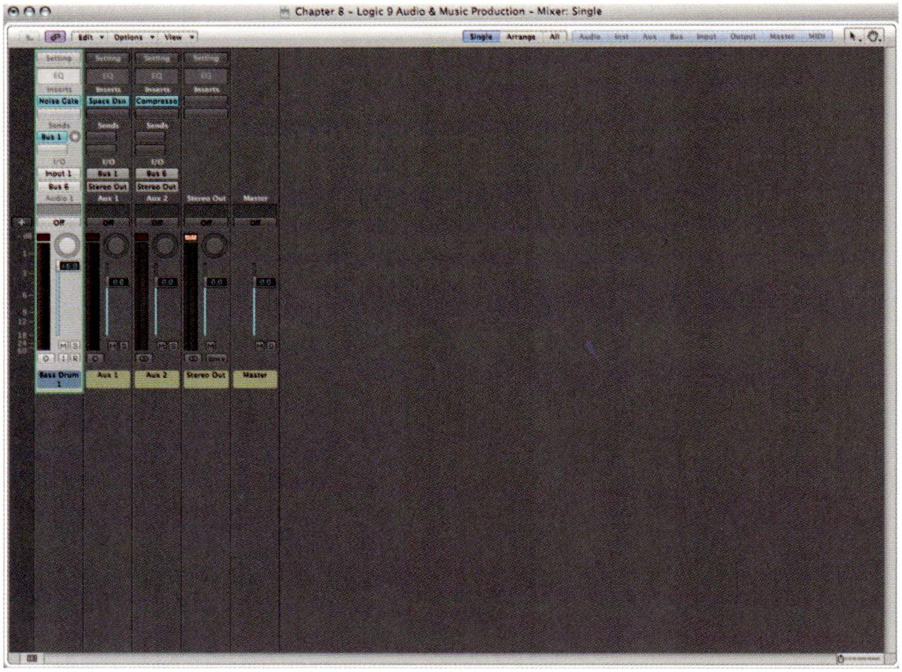

Figure 8.14 In Single mode, you can see the complete signal path for the currently selected track.

8.4 Folders and the Mixer

As we saw in Chapter 4, folders can be useful way of packing your arrangement into groups of related tracks, which can then be viewed and edited independently of the main arrangement. For example, you might decide to pack away your drum tracks into an individual folder, keeping those tracks away from the main view of the track.

Not surprisingly, therefore, the folder feature also has an interesting impact on your view of the Mixer. Any folder tracks are displayed as small strips in the Mixer, without any form of level control, mute switching, and so on. However, by clicking on the small folder, or entering the folder via the Arrange page, you can access this sectioned off part of the mix, using the small hierarchy tab in the top left-hand corner of the Mixer to move backward into the main Mixer.

Managed carefully, folders can be a useful visual tool for keeping on track of the sections of a mix, although, as we'll see later on, there are also other functions for the control of the groups of instrumentation from a sonic perspective. One of the tricks is to decide what you keep on the top level of the Arrange area. For example, if you delete the track lanes from the top "arrangement" level, they'll only be viewable once you've moved into the folder. This can be a great solution for hiding parts of the mix away, although not so good if you still

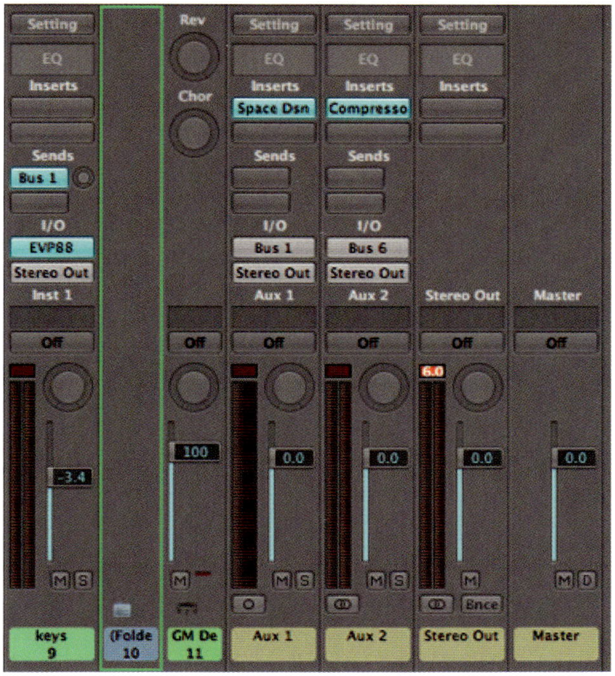

Figure 8.15 Use the folder function to "pack away" parts of the mix. This can be useful in projects with large amounts of channels or tracks.

want to keep an eye on the entirety of instrumentation. As an alternative, retain the track lanes in the arrangement (even though the regions have been packed away into a folder), only using the folder functionality to "hone in" on specific parts of the mix as and when you want to.

Plug-In Boxout 2

Helpers … Handy Little Plug-Ins

Sometimes the small, innocuous little plug-ins can be as vital to creating an effective mix as the big processor-hungry reverbs. Take the Gain plug-in, for example, its main application is simply to raise the gain of an input – surely not that useful then? Say you've recorded in a series of complicated automation moves but simply want to raise the entirety of the track by 2 dB. Usually this would necessitate editing or rerecording the moves. Alternatively, you could use the Gain plug-in (patched somewhere in the insert path) and simply select a 2 dB boost! Some of its other applications can also be great on stereo signals – switching the left- and right-hand sides, for example, or mono'ing a stereo mix (try leaving an instance across the stereo bus) to check its mono compatibility. Also, if you need to phase invert any microphones (a snare bottom, for example), simply activate the Phase Invert option.

The Multimeter plug-in is a great addition to any stereo bus, providing up-to-date information on the track's spectral properties (using a 1/3 Octave Spectrum Analyzer), alongside phase characteristics courtesy of a distinctive Goniometer. Use the spectrum analyzer to get a better grasp

Figure 8.16 Many of Logic's smaller plug-ins can aid the effective workflow of a mix – from the Gain plug-in to the range of metering options.

on the spread and range of frequencies in your mix. If you're using monitors with a limited bass response, the lower spectrum information (100 Hz and below) can be crucial to keeping an eye on any excessive activity. The Goniometer provides some indication of the stereophonic properties of your mix – a strong mono image is indicated by a clear line up the middle of display, while excursions into stereo are indicated by movement to the left and right.

The Correlation Meter – available as a separate plug-in or, on either of the Goniometer or Analyzer's screens – indicates phase problems with the mix. If the phase is good, you should see consistent movement one side of the line (usually in the 11 region). However, if the phase is bad (indicating problematic mono compatibility), the Correlation Meter moves between both –21 and 11 in a continuous fashion.

8.5 Beginning a Mix

With the mechanics of understanding how Logic's Mixer works under your belt, let's start to have a look at the creative and technical processes of putting the mix together. However, before you start instantiating various plug-ins, take time to establish a plan for what you want to achieve. Think about how you intend to distribute instruments across the soundstage – in respect to both left-to-right placement (pan) and front to rear (with the use of reverb and maybe EQ). Establish what you feel should be the "lead" instrumentation – in other words, the three or four instruments that really hold the track together – alongside the other elements that should only provide a "supportive" role. Finally, identify instruments that are possibly fighting against one another, sounds that may blur the definition and energy of the track, or parts that may get lost among other instrumentation. Ultimately, if you're clear about some of these issues from the start, you'll have a much better chance of producing a coherent end result.

Basic Part Leveling

An essential part of the mix revolves around good leveling practice – put simply, balancing parts without the addition of any additional processing. Although the precise technique varies from engineer to engineer, the intention is to get a good "working balance" of sounds, some with more dominant roles than others, but all without overloading the main outputs from Logic. For example, it's easy to start putting a mix together, pushing channels into the 13–16 dB region only to discover that the main output quickly starts to distort. Instead, try to keep an eye on the main output level (usually on the output 1–2 channel, which should sit side-by-side the instrument channel/audio channel in the inspector) and bring in two or three of the key instruments leaving 6–12 dB or so of headroom. Unless an instrument has been recorded particularly "hot," this will probably be in the region of –24 to 0 dB.

Alongside the basic of part leveling, you might also want to consider panning the various channels to create a realistic soundstage. Think about the instrumentation arranged across the stage, creating a natural, well-distributed configuration of the instrumentation, without relying too much on the extremes of left and right (–64 and +63, respectively) and leaving a clearly defined space in the center of the mix for important lead instrumentation like vocals (in other words, only a few channels should be left in the 0 position). Remember that as you apply effects like reverb and delay (which we'll see more of later), you can still pan these signals just like any other channel in the mix.

Applying Equalization and Compression

Two of the most important signal processors in Logic – and the cornerstone of any professional mix – have to be the equalization and compression plug-ins. One of the main applications of compression and equalization is to provide better separation between sounds – fixing different elements into specific parts of the frequency range of a track, for example, or locking a sound better into the mix's dynamic properties. One common technique with EQ is to reduce clashing frequencies between sounds, allowing each sound to sit in its own sonic space. For example, a 200-Hz reduction on a low guitar part would facilitate a little more space for a bass, whereas the intelligibility of the vocal could be improved by a cut at 1–3 kHz on any competing instruments. Compression, on the other hand, helps iron out any dynamic inconsistencies – for example, rather than a vocal dipping in and out of the mix, it can remain consistently loud.

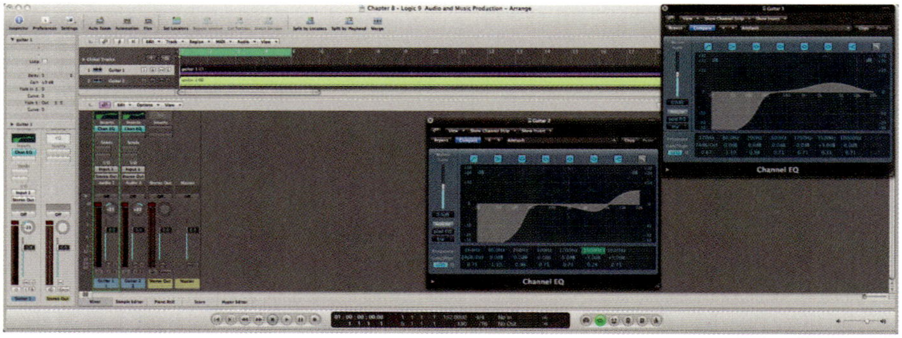

Figure 8.17 Try using equalization to create better separation between instruments. In this example, a 3-dB boost at 3.5 kHz in one guitar has been balanced out by a cut in 3 dB at 3.5 kHz on another.

Plug-In Boxout 3

Compressor

Along with EQ, compression forms one of the fundamental tools of a mix-down and essentially works as a form of automated amplitude control – the compressor reacts to loud peaks in the input and then attenuates the level of the signal accordingly. The two key parameters in this process are threshold – rated in decibels and governing the point at which compression begins – and ratio, which sets the hardness of compression applied. For example, with a low threshold and ratio (−220 dB and 2:1, respectively), a slight compression would be applied, even on relatively quiet parts of the signal. Alternatively, a higher threshold and ratio (−25 dB and 10:1, respectively) would produce a stronger compression, but this would only be applied on the rare occasion the signal level exceeded −25 dB.

In addition to the key compressor settings (threshold and ratio), Logic's compressor plug-in contains a number of other parameters to further refine the processor. Attack and Release, for example, are important tools for defining how the compressor moves in and out of gain reduction, as the signal exceeds the threshold. With a quick Attack and Release setting, the

Figure 8.18 Logic's compressor is a versatile gain reduction tool, capable of controlling the precise amplitude and dynamic range of a signal in the mix.

(Continued)

compressor is set in a "fast-acting" mode – the response to loud transients (like loud snare hits) is quick and efficient, and the compressor is quick to return to its normal state once the input reduces below the threshold. As great as this sounds, however, these fast movements in and out of gain reduction can have a negative effect on sound being processed, as the compressor creates a distracting "pumping" sound. Setting a slower Attack and Release might allow the odd loud peak to slip through the net, but you'll end up with a more musical, empathetic response from your compressor.

Peak and root mean square (RMS) modes govern how the compressor "listens" to its input. Essentially, the Peak mode responds exactly to the true level of the input, while RMS responds to an averaged level, closer to how the ear perceives loudness. With a Peak detection setting, therefore, you'll find the compressor reacting more than on the corresponding RMS setting, although its response might be considered slightly less musical. The Knee of the compressor is a useful means of creating a more graduated transition on harder ratio settings – the ratio is slightly softer (2:1, for example) ahead of the threshold, only reaching its full strength (6:1) a few decibels after the threshold.

As the overall effect of compression is to reduce the dynamic range of your input (in other words, making the loud bits quieter), you'll probably find your corresponding output quieter than without compression. Increasing the output Gain (at the right of the interface) restores levels lost through compression. You can also use the AutoGain option of Logic to apply this automatically, although in some cases this can produce distortion.

Compression and EQ both can be applied via the insert path of a selected audio channel strips, instrument channel strips, or aux channel strips, although the EQ can also be quickly activated by double-clicking in the small EQ box at the top of the channel strip. Double-clicking the selected plug-in will then open a floating dialog, allowing you to adjust the various parameters contained within the plug-in. Of course, bypassing effects is essential to establishing exactly what you have or haven't achieved. This can be done using the dedicated bypass control in the top left-hand corner of the plug-in's interface or directly from the Mixer itself by option-clicking on the appropriate plug-in slot.

Adding in Further Plug-Ins

Further plug-ins can also be inserted as part of the path of plug-ins you create on the channel strip. Remember, though, that the order in which you insert the plug-ins can have a big effect on the overall treatment produced. Even something as simple as EQ and compression, for example, can have a subtly different

Figure 8.19 The order in which plug-ins are inserted can have a big effect on the end output. Try using the Hand tool to experiment with the order of your plug-ins the channel strip.

output based on which plug-in is placed first in the chain. More extreme effects, say, for example, distortion followed by EQ or EQ followed by distortion, can have completely different result given their relative position. If you need to change this order, you can use the Hand tool on the Mixer (or simply press Cmd + click the plug in) to reorder the inserts in any way you see fit.

One common mistake at this point is to apply too many plug-ins, possibly in a desperate attempt to improve apparent deficiencies in the source recording; "at least if it's covered in distortion and delay, nobody will notice that it's out of tune!" Try to have a clear strategy in your application of effects – not all tracks necessitate the use of plug-ins, and where a more complicated series of plug-ins are used, try to make this a unique and identifiable "special" feature, rather than the norm.

8.6 Adding Send Effects

As tempting as it is to patch everything across a channel's insert points, it's really worth making a clear distinction between the use of true insert-based effects and the use of send effects. The use of an insert effect implies there's no need to balance an unprocessed version and a processed version of the signal. EQ, therefore, is a perfect example of insert effect as there is little or no need to hear both an unequalized version and an equalized version of the sound at the same time. The same could also be said for compression. Reverb, on the other hand, is a clear candidate for being applied as a send effect, with the need to balance the respective levels of wet (in other words, reverberated) signal with the dry, unprocessed version. In theory, the greater the reverb, the more the channel appears to move toward the back of the mix.

Plug-In Boxout 4

Channel EQ

The Channel EQ is easily one of the most important plug-ins in any mix, controlling the precise spectral qualities of signals passing through it. Although the Channel EQ takes a low drain on available CPU resources, it is both a powerful and a flexible tool, split into eight frequency bands with a cut/boost and frequency parameter for each band. As a good starting point to understanding your input source, try activating the Analyzer feature of the EQ. The Analyzer provides a real-time Fast Fourier Analysis of your signal, indicating the distribution of energy across the sound spectrum – a bass, for example, should produce large "humps" formed by its fundamental at 100 Hz, alongside additional harmonics further up the harmonic spectrum.

The two extreme bands – at the far right- and left-hand side of the interface – govern the controls for the high-pass and low-pass filters within the EQ. Effectively, these completely remove frequencies above or below the given cutoff point – try using the high-pass filter, for example, to tame any excessive low-frequency energy. Moving inward, the next two bands correspond to the shelving EQs, which are somewhat comparable to the treble and bass controls on a conventional hi-fi. Use the shelving EQ for general sweetening activities, creating the familiar "smiling" EQ curve with a boost at around 80 Hz and 12 kHz, respectively.

The remaining bands work as traditional parametric EQs, with a cut/boost control, frequency setting, and a fully variable Q parameter. Q sets the width of cut and boost, and therefore the resultant amount of cut and boost to harmonics is near to the EQ's selected frequency. A wide Q is a useful way of shaping more "general" qualities of the sound – maybe a lack

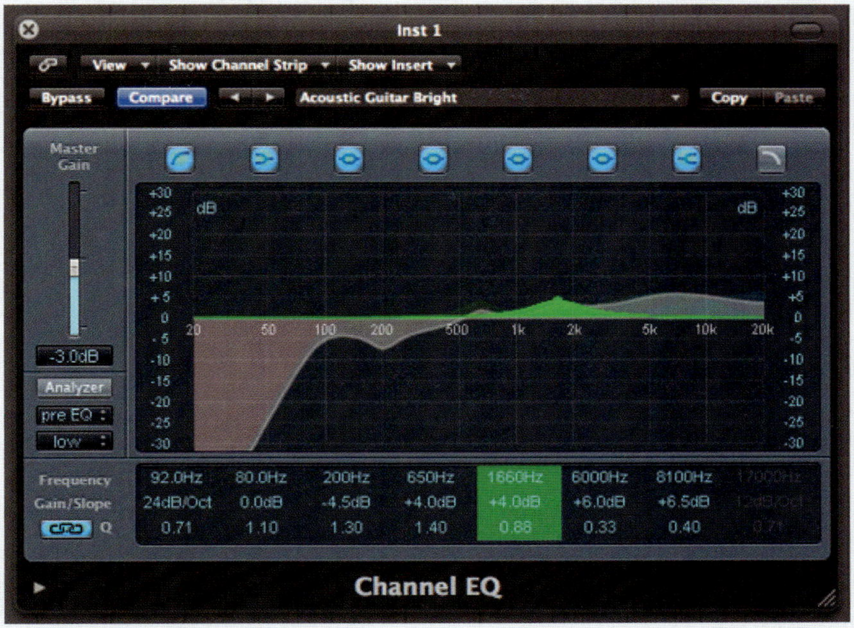

Figure 8.20 Channel EQ is one of Logic's primary mixing tools: with four fully parametric bands, two shelving controls, and two filter sections.

of bite in the upper mids, for example, or an overall woolliness in the low mids. Use a tighter Q where your frequency issues are more specific – like a boomy resonance on an acoustic guitar, for example, or prominent problem harmonics.

The first step to creating a send from any given channel is to double-click in one of the available send slots on the strip and select from one of the 64 available bus destinations. Raising the level on the small send pot will then bleed an amount of the channel's signal via the bus to the appropriate aux channel strip. The aux channel strip should then have the corresponding plug-in (an instance of Space Designer, for example) patched across its insert path so that signals entering the aux channel strip are affected accordingly. With the mix parameter on the plug-in set to 100% wet, you can now control the level of reverb fed back into the mix, using the aux channel strip's fader or using further plug-ins across the aux channel strip to further process the sound of the reverb.

Both reverb and delay are important tools in defining the spatiality of your mix – in other words, the front-to-back perspective of the soundstage. Having worked carefully on your level and panning, you should have initially formed a good representation of your left-to-right soundstage. However, with the application of delays and reverb, you finally have a chance to set sounds forward and backward in the mix – a dry vocal, for example, sat squarely at the front, or a distant keyboard pad drifting toward the back of the soundstage. One really useful technique is to place two to three contrasting instances of Space Designer to establish a front, mid, and rear acoustic to in your mix. Use some of the shorter reverbs on close sounds, or rhythmic elements (like drums or rhythm guitar), that don't suit long reverb settings. The longer settings, on the other hand, could be reserved for a few unique sounds that really benefit for longer reverb tail and a more distant mix placement.

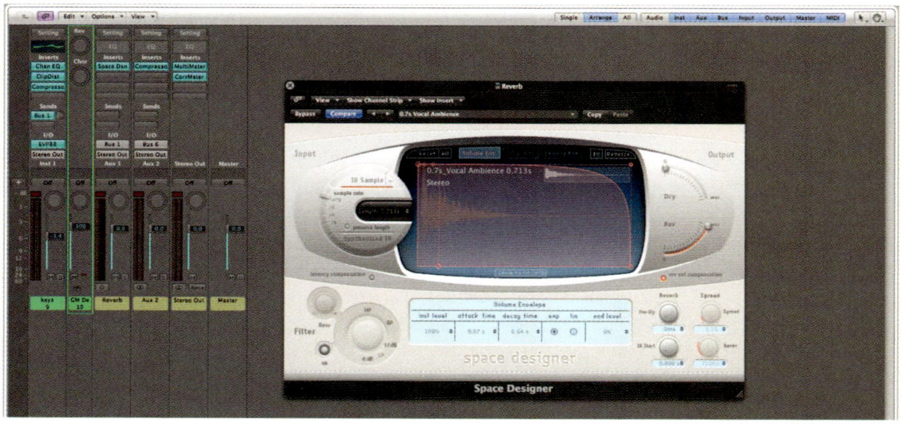

Figure 8.21 A basic configuration of send effects – the keys channel has been set up to have an auxiliary send delivering a little of its signal through to an auxiliary channel which is configured with a reverb on its 100% wet setting.

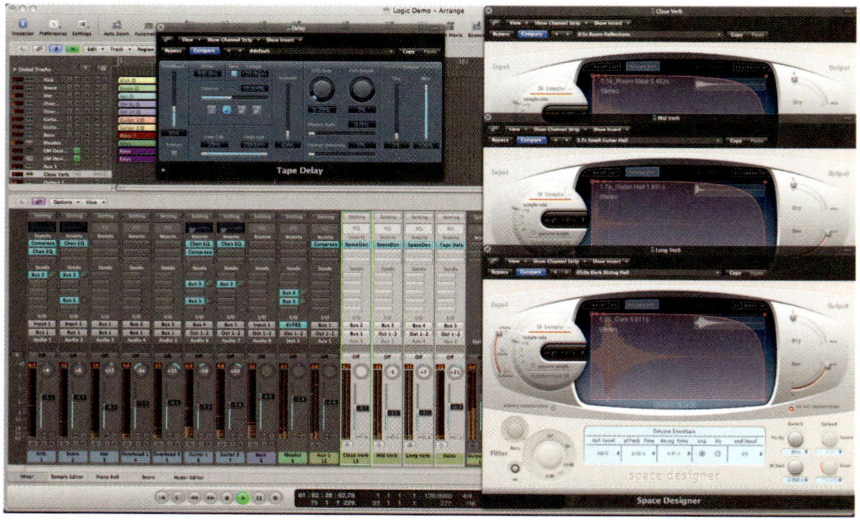

Figure 8.22 Use two or three complementary reverbs to help define the front-to-back perspective in your mix. Additional delays can also help define the "spatial" dimension.

One important concept of the sends is whether they are configured for either pre or post fade operation. In a mix, the most common example (and the one Logic defaults to) is to use the sends in post fade mode – that is, the send happens after any adjustment in level for the particular channel. If, for example, a fader level is brought down in the mix, its corresponding amount of reverb is also attenuated – in effect, preserving the ratio between the two sounds (dry and reverberated). Pre fade sends (as we saw in Chapter 4) are usually associated in the creation of headphone or cue mix. To adjust between pre fade, post fade, and post pan operations, simply click and hold on the send and adjust accordingly.

8.7 Combined Processing Using Aux Channels

Having seen one application for aux channel strips – that of applying send effects – let's have a look at the other ways in which they can be applied during a mixdown. By changing the output option of any audio or instrument channel in the mix, you can also decide to route it to an aux channel strip ahead of the signal reaching any of the main, physical outputs from Logic. Used in this way, you can combine a number of signals – say, the various microphones positioned around a drum kit, for example – to a single fader, with the option to control both its level and the application of additional processing en masse.

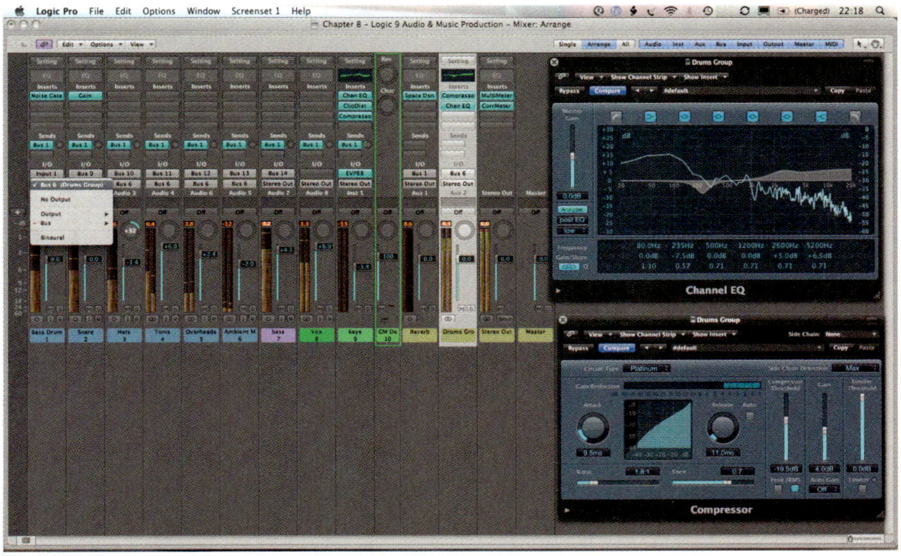

Figure 8.23 Bus processing can be achieved by sending a number of channels (via a bus) to an aux channel. On the aux channel, insert the required equalization or compression, for example, so as to process the "group" of sounds.

The use of so-called bus processing across aux channel strips can be a great way of locking together the principal components of a mix, as well as helping groups of sounds gel in an effective way. Many American rock engineers get a great deal of mileage from this technique, combining compression on a channel-by-channel basis, alongside compression across a collection of aux channel strips. The result is a mix that "pumps" in an empathetic way to the source material, providing a real intensity and sense of loudness brought about through the reduction of dynamic range.

However, if you do intend to apply a lot of compression on your mix, it's well worth making sure that you use a variety of different compression techniques as a means of forging the identity between different sounds (rather than the track just sounding squashed), as well as matching the style of compression to the instrumentation you're processing. Try using low ratios (1.5:1 to 2:1) and low thresholds as a means of massaging the sounds into place, whereas harder ratios and higher thresholds (6:1, −5 dB) could be a great way of simply slicing a few loud transients of a drum submix. With any bus compression, you'll also need to pay close attention to the Attack and Release times of the compressor, with slower settings (especially on the release) providing a more musical result.

Plug-In Boxout 5

Convolution Reverb and Space Designer

Convolution has quickly become the accepted standard in producing professional-grade reverb treatments. The technique works by taking an acoustic sample, known as an impulse response or IR, from a given room or indeed, a hardware reverb unit. The process of convolution then simply takes the short IR file and mathematically "folds it over" the source signal, effectively recreating the effect as if the sound had been recorded in the same space or through the same reverb processor. For that reason, convolution reverb can sound astonishingly realistic, although this can be at the expense of your available DSP resources. Ultimately therefore, it makes sense to run just a few instance of Space Designer, patched across bus faders, than lavish multiple instances on individual channel faders.

As part of the Logic Pro install, Space Designer comes with an impressive Library of prerecorded impulse response files (stored under Library/Application Support/Apple/Impulse Responses). You can load these simply by scrolling through the presets or loading them in via the Load IR option to the right of the IR sample switch. Additionally, you could choose to make use of one of the growing number of third-party IR libraries, like Spirit Canyon's Spectral Relativity, or indeed, your own custom-sampled IR files.

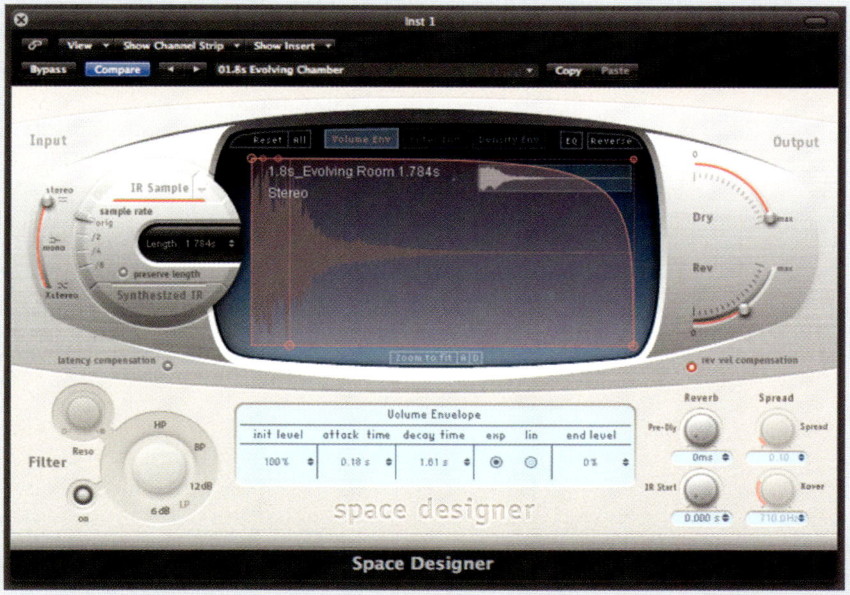

Figure 8.24 Based on unique samples of the original spaces, Space Designer's convolution reverb engine produces some stunningly realistic reverb treatments.

Like a growing number of the professional convolution reverbs, Space Designer offers a surprising amount of control and flexibility with effects you can achieve with it. Once an IR sample is loaded, you can immediately adjust both its volume envelope and its filter characteristics, assuming the filter is activated. Besides being able to produce some subtle modifications (rolling off the high-end, for example, with a touch of low-pass filtering), these tools can be great for abstract ambience treatments – try using the reverse setting and some extreme filter movements with lots of resonance! If you need to shorten the reverb tail, adjust the length parameter. Alternatively, to create a longer, darker reverb, adjust the sample rate into one of its slower settings. This can also be a great way of producing longer reverb times, without maxing-out your DSP resources.

8.8 Using Groups

As alternative to bussing sounds together via aux channel strips, you can also make use of Logic's Group feature – as we've already introduced, with the concept of edit groups in Chapter 5. Taking the concept further, we can also use Groups to lock together collections of channels, say a group of drums, for example, or backing vocals. But given the aux channel strip bussing system, what use is an additional means of ganging faders?

As simple as the bussing system is, there's one major conceptual problem – that of post fade send effects like reverb. Imagine a collection of kit sounds all being sent to an aux channel strip, with a snare also making use of a generic reverb (also shared by the vocals and strings) on send 8. If the level of the kit were then reduced on the aux channel strip fader, you'd hope the reverb level would also be reduced accordingly, but this is not the case. As the send happens before the aux channel strip, any subsequent level changes will have no effect on the reverb – effectively the snare will get "wetter" as the level of the kit falls. Of course, one simple solution to this would be to route the reverb to the same aux channel strip, but this would then have a knock-on effect on the vocal and strings reverb!

So, here we see a perfect example of the application and benefit of grouping channels as apposed to bussing signal via aux channel strips. Grouping retains all the important gain structures of the original mix – the ratios of reverb are retained, even as the level of the group is reduced. Bussing still has its place, though, as there's no way to apply effects like compression and EQ onto the group. So, in theory, a good mix may well involve combination of both grouping and bussing via aux channel strips, grouping to "lock" collections of faders, and bussing via aux channel strips to provide grouped applications of effects!

Looking more closely at the Group Settings dialog, you can see a number of additional features that could potentially improve the speed and efficiency of your mix. Besides obvious grouping controls like volume and mute, you can also link the send controls (this could even be helpful in quickly modifying cue mixes as you're recording), as well as any linked Automation Mode selections.

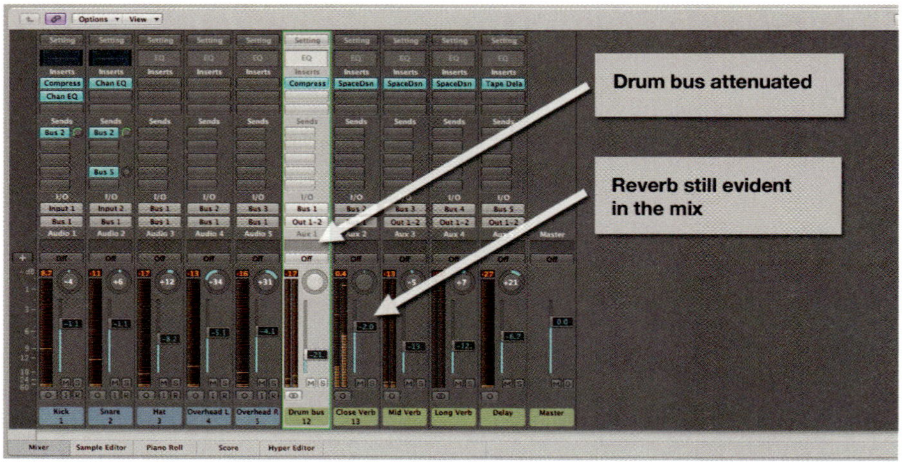

Figure 8.25 Shared reverbs can make the use of aux channels for grouping purposes problematic; although the fader level of the drums group is reduced, the reverb level stays the same.

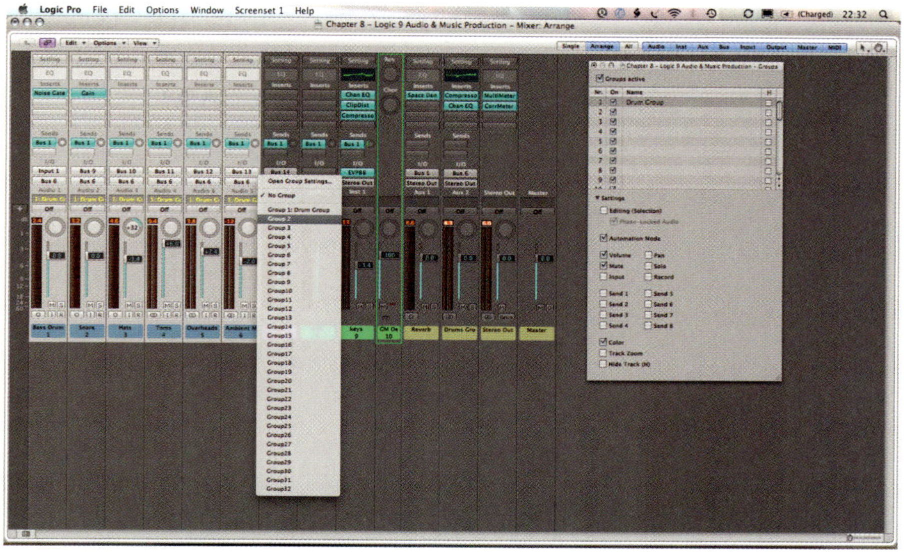

Figure 8.26 This example uses both groups and bus processing. The Group allows us to better control the level of the drums in the mix, while the aux channel is solely used for processing rather than level control.

However, one vital keyboard shortcut to use in conjunction with the Groups is the Toggle Group Clutch command. By default, this is configured onto Cmd+G and can be used to temporarily disable all groups – a great way of quickly modifying an "internal" balance within any number of groups, without having to laboriously enable or disable the relevant groups.

Knowledgebase 2

Parallel Compression

Parallel compression has become a real buzzword in mixing circles, but what is it, and can Logic be realistically used to apply it? Unlike the traditional insert-based approach to applying compression, which works on the assumption that you'll only want to hear the 100% compressed signal, parallel compression offers the unique possibility of hearing both compressed and uncompressed versions. Although parallel compression is arguably less effective at reducing the dynamic range of a signal, it is an excellent way of combining both the hyped-up sound of heavy compression, with the natural dynamics and "air" of an uncompressed recording – yes, to use the cliché, it really is the best of both worlds!

One really simple way of creating parallel compression is to copy the source track onto two adjacent channels. On the one channel, keep the signal

(Continued)

uncompressed and open, while on the other, try setting up really juicy, pumping compression. Don't worry about the integrity of the signal – pick a suitably tough ratio (6:1 or harder!), relatively quick Attack and Release, and let gain reduction pump well into the 6–10 dB range. Some engineers will even EQ this channel, picking out the extreme highs and lows (80 Hz and 12 kHz), as well as applying a little tuck in the mids. With the compression set up, combine the two channels together, mixing in the compressed version to add "balls" and body to the uncompressed track.

Another interesting approach is to use the compressor almost like a send effect – blending any number of sounds through to the same heavy compression setting. Try doing this on a drum mix, sending mainly the close snare and kick drum through to the paralleled compression. Again, pick some suitably pumpy settings, maybe even a touch of extra ambience, and sit the bus in the mix to add the required amount of weight and importance.

Figure 8.27 In this example of parallel compression, a bus send is being used to add compression using the same routing technique as a reverb send. Use a heavy compression setting, and then bleed in the required amount compressed signal to add "body" without compromising transient detail.

8.9 Working with Channel Strip Settings

As you're building up the mix, you may well find that certain channels require similar settings to that of others – maybe you've compressed and equalized one backing vocal, and you want to apply the same setting to corresponding

Figure 8.28 Use the channel strip settings as a quick way of moving one channel's plug-in configuration to another. The presets feature also allows us to do this between songs.

backing vocal track. Clicking on the small arrow to the right of the Insert label allows you to open up the channel strip settings menu. Try using the copy/paste option as a simple means of duplicating the channel settings on a number of channels. Alternatively, if you have a number of favorite channel strip settings – maybe a particular compression and EQ on a given vocalist, for example – you can also save the presets off, as you would any of the other plug-in presets. These favorites could then be recalled at any point on any given song.

8.10 Automation: The Basics

Despite the considered application of compression, you may still find the demands of your mix changing from verse to chorus and vice versa. Ultimately, although one balance may work at a given point in time of the song's development, the same might not be true as further instrumentation enters and the musical qualities of the track change. In the "golden days" of recording – before computers stepped out of the office and into the studio – it wasn't uncommon for a mix to involve "all-hands-on-deck": the engineer, assistant engineer, and even the members of the band, all frantically pushing faders up and down to shape the mix to the dynamic of the song.

Logic's Automation provides complete control over qualities of the mix throughout the duration of the song – including basics like volume, pan, and mute and more advanced options like plug-in parameter automation. The golden rule to remember with automation is only to apply it toward the ends of a mix – once you've established the principal balance, equalization, compression, and effects usage, only then should you begin to turn to automation. The problem is that any mix changes after automation, although not impossible, can be quite a headache to perform – potentially requiring you to rewrite or delete existing automation moves.

8.11 Track-Based versus Region-Based

One potentially confusing aspect of automation is that Logic contains two contrasting methods for applying automation – track-based automation and region-based automation. In truth, region-based automation – where automation data is contained within the audio or MIDI region itself – is actually a throwback to the earlier versions of Logic and has largely been retained for the purpose of backward compatibility, although, to be fair, it still has its uses. Track-based automation – where automation data is recorded on a separate multiple track lanes to that of the audio region – offers a far more flexible means of creating and editing automation data, as well has having the distinct operational benefit of not being tied to the regions in questions. Editing or moving a guitar solo, for example, won't necessarily disrupt the automation data that accompanies it.

Thankfully, Logic does provide a means of transporting one automation type (region-based or track-based) to another – making the combination of the two

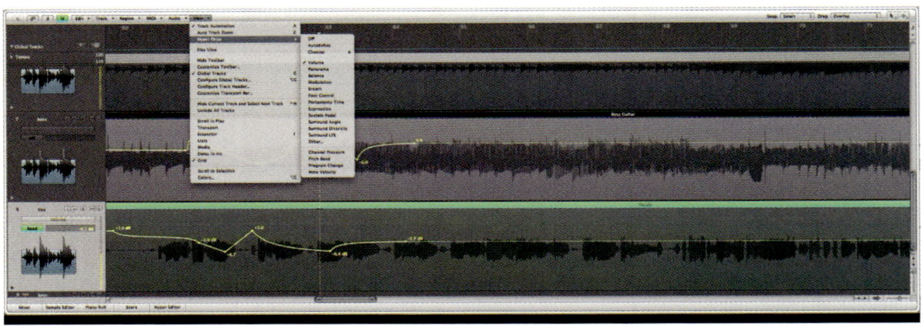

Figure 8.29 Two different approaches to automation – the old-fashioned Hyper Draw (top) and the standard track-based automation (bottom).

approaches quite a powerful feature. For example, you could write the automation using the superior track-based controls and then convert this information into automation data that is stored with the region. For now though, we'll take a look at how track-based automation works and then consider the region-based case later on.

Plug-In Boxout 6

PlatinumVerb

PlatinumVerb is a valuable alternative to Space Designer, either for Logic Express users (who don't have access to Space Designer) or for situations where you're running slightly low on available DSP resources. In contrast to Space Designer, PlatinumVerb places a surprisingly low drain on the CPU, although correspondingly, you may find its output slightly less authentic than Space Designer. Used carefully however – on short settings with drums, for example, or any instrumentation sitting lower down in the mix – it can produce surprisingly effective results.

Figure 8.30 If you're short of processing resources, or don't have access to Space Designer, PlatinumVerb makes for an excellent alternative source of reverb.

(Continued)

One of the best things about "modeled" reverb, as apposed to convolution reverb, is that it offers precise control over the sound produced in the virtual room. In the case of PlatinumVerb, it divides the reverb up into two key stages – the early reflections, as sound initially bounces back from the walls, and the more diffuse reverb tail, as the reflections merge together to create a definable "trailing-off" to the sound. Try moving the Balance slider over each side to fully understand and audition the effects obtained in each stage.

Using a predominant mix of early reflections can be a great way of thickening up drum sounds by producing a noticeable, distracting reverb tails. Experimenting with the room shape, stereo base, and room size will change the qualities of reflections produced, either by creating a tight centralized reverb or by creating a wider, more expansive set of reflections. When it comes to the reverb tail, try exploring the density, diffusion, and reverb time parameters. On the whole, most conventional reverb effects tend to stay within the realms of 1–2 s, and, not surprisingly, this is where PlatinumVerb seems to sound at its best. Density and diffusion, respectively, govern the spacing and randomness of the taps. Try using both in their lower setting for an effect like Spring Reverb (great on electric guitars), or use a higher setting for a smoother sounding reverb.

8.12 Automation Modes

To engage a channel into automation, you'll need to change automation mode according to the way in which you intend to write automation data into Logic. By default, all faders are set to their Off position, just above the pan pot – this means that any existing automation data are ignored and the faders can be freely moved or repositioned (without fear of Logic snapping them back!) at any point. Engaging any of your faders into an automation Write mode (Write, Touch, or Latch) will allow you to begin writing data based on the current song position. Interestingly, though, the "recording" of automation data is independent of the transport's record switch; in other words, you only need to be in Play mode for automation to be written.

First, let's take a look at the various modes used to record automation data.

Write

Think of this mode as the most dangerous! Any fader engaged into write mode will record data onto the automation lanes, even if there's existing data, so this should really be used with care. Write, however, can be a valid way of deleting and replacing automation data in one pass.

Touch

Touch is the safer way of writing automation data, as it only engages into writing data once the fader has been "touched." Release the fader – even as the track is still playing – and the fader will return the previous recorded position, and carry on reading any existing automation data. Touch, therefore, can be great in creating a few strategic nips and tucks in your mix – briefly either lifting a phrase out for a few seconds or pulling back any part that dominates the mix.

Latch

Latch mode is comparable to touch, in that a fader is only engaged into writing data once it has been touched. However, when the fader is released, it continues to write data – potentially erasing any existing moves – and the level it was last left at. Latch can be a useful mode when you need to raise or lower a level of a part and then leave it at that level for the remainder of the song without having to constantly "hold" the fader in place. Stopping the transport will, of course, stop the writing of automation data.

Read

Read is a "safe" automation mode, where any automation data is read back but no further moves can be written – indeed, if the fader is moved for any reason, it will promptly snap back into place! On the whole, most users tend to leave faders in Touch mode as they're automating, allowing them quickly engage the writing of automation data without having to constantly switch modes. However, to reduce the risk of overwriting certain aspects of the mix, it's best to switch all faders back to Read mode once you have finished automation.

Although the description of different modes has assumed you're working with the channels faders, it's important to remember that once a channel has been placed into Write, Touch, or Latch, any movement in its accompanying parameters – including mute, pan, or any of the plug-in parameters – will also

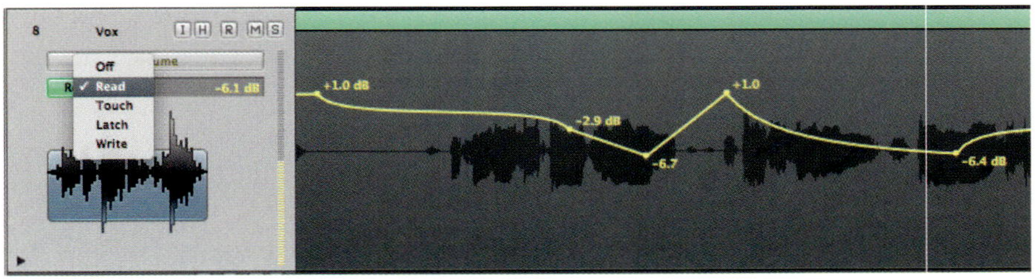

Figure 8.31 Each of the four different automation modes used in Logic has its own impact on how you read and write automation data.

be recorded. This allows any part of the mix to be automated with the same degree of flexibility as the channel faders; for example, the reverb time of a vocal reverb could be swelled going into a "larger" chorus, or the feedback of a tape delay unit could be modulated to create some dub-inspired delay treatments on the end of a line.

Plug-In Boxout 7

Using Delay

Although most users tend to perceive reverb as the main tool for defining spatial qualities of a mix, it's also surprising to realize just how much spatial interest can be added with something as simple as a delay line. Logic comes with three different delay plug-ins – sample delay, tape delay, and stereo delay – facilitating a range of different delay treatments.

As the name suggests, tape delay is modeled on the type of delay effect produced by classic tape-based delay effects like Roland's Space Echo or the WEM CopyCat delay. As these processors use tape to produce their delays, the effect has a characteristic "dark and dirty" sound to it. By default, the plug-in works with tempo divisions – simply select the required division (semibreve, crotchet, quaver, and so on) and the delay will appear in time with your track, even if the song's tempo changes. Use the Groove slider in the extreme settings (33% and 75%) to change the division to dotted note values. You can also achieve some great slap-back

Figure 8.32 Logic's simple delay plug-ins can be a surprisingly effective tool for mixing – from rhythmic delay effects to subtle forms of "slap-back" ambience.

delays (an effect famously used on early rock'n'roll vocals) by using a semiquaver setting and sliding the Groove parameter down into the region of 33%–50%.

Feedback sends a proportion of the sound back on itself, effectively creating a regenerative delay effect. On the tape delay, feedback settings in the region of 50%–100% will appear to "hold" the delay, with an increasing amount of distortion and grit on each repeat. Stereo delay works on the same principle as tape delay, although it deliberately avoids the tape-based coloration, in preference of a cleaner digital repeat. You can also set different delay time for the left- and right-hand side of the stereo image.

In contrast to the other delay plug-ins, sample delay only deals with incredibly small delays times, measured in samples. As a rough guide, 44 samples equate to 1 ms of delay, so even at its maximum setting (4000 samples), sample delay only gives us about 90 ms to play with! So why use this plug-in? Well, sample delay can be beneficial in applications fixing minute time delays caused when microphones are widely space apart – applying sample delay to correct these anomalies and create a more phase-coherent image. If you're less technically inclined, try using a small amount of sample delay on some drum room microphones to simulate the effect of sound reflecting further away from the kit.

8.13 Viewing and Editing Automation

By default, Logic hides the display of automation data – otherwise an arrangement could soon become cluttered. However, once you've started recording a few moves, you might want to see how the mix is beginning to shape up. Select View > Track Automation to display the current recorded automation – you might also want to toggle this using a keyboard shortcut (the default is A) so that you can quickly switch automation viewing on and off.

With the automation view engaged, you'll notice some important changes to both the tracks and the arrangement itself. Looking first at the tracks, you should now see the automation mode indicated on them (Read, Touch, and so on) alongside the current viewable automation parameter – clicking on this should allow you to scroll through all the parameters available for automation. You'll also see a small bar graph meter indicating the current fader position. Note that this can be freely modified and controlled, just like the "real" fader as part of the Mixer, and can be a great way of adding in a few cunning automation moves.

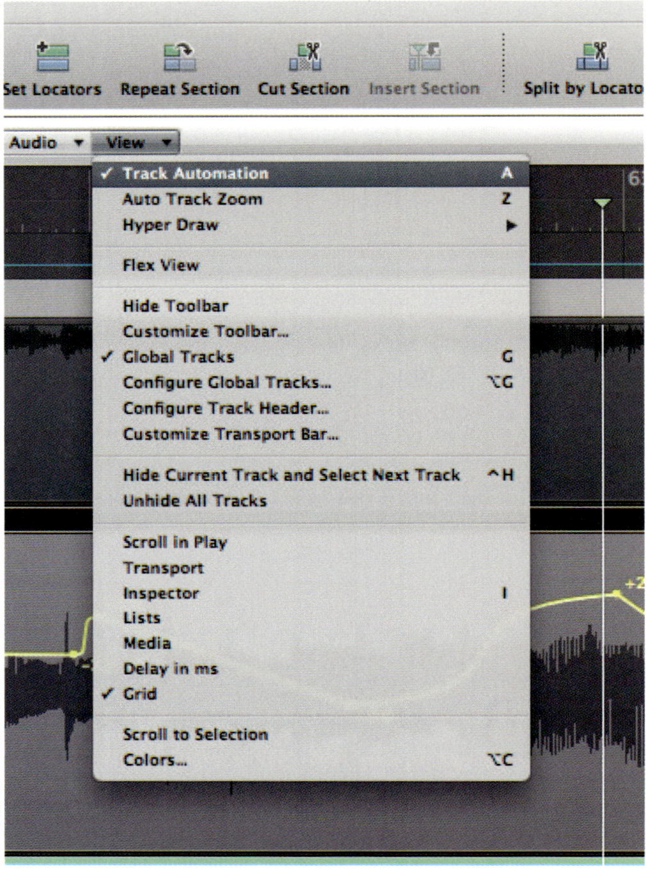

Figure 8.33
You'll need to enable the automation view mode to see or edit the moves you've recorded.

Where multiple plug-in parameters have been automated, you can also add further track lanes into the equation, simply by clicking the small arrow toward the bottom right-hand corner of the track name in the track list. Alternatively, option-clicking opens as many automation lanes as is required to display all the automation data currently recorded.

Automation data itself is displayed as a series of nodes, which can be manipulated using the usual Pen, Eraser, and Arrow tools. As you'd expect, the Pen tool allows to draw new automation moves, while the Arrow can modify existing node points, or by double-clicking add new nodes into the equation. Indeed, in many situations, it may be quicker and easier to draw in a couple of automation nodes than to laboriously write in a series of moves. Manually placing the nodes can also be a good way of setting in long automation events – a long filter sweep over 30 bars, for example – or when events, like fade-outs, need to happen at a precise time.

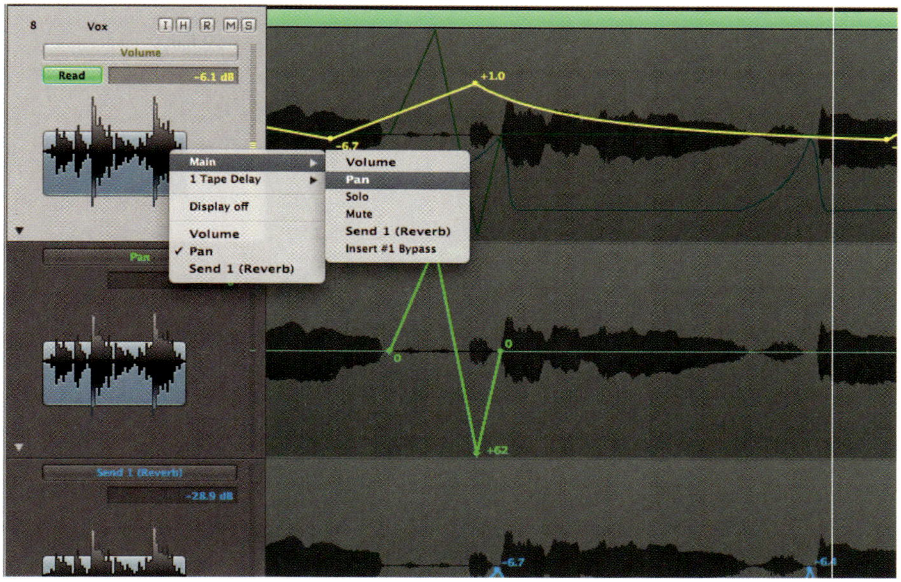

Figure 8.34 Press the small arrow on each track lane to open up a parallel automation lane, making it clear to see the moves on several different parameters on the same track.

Plug-In Boxout 8

Vocal Processing

Vocals can be one of the trickiest parts of a mix to get right, so it's great to see a range of plug-ins suitable for vocal processing in Logic. Away from the core plug-ins that we've already covered – including EQ, compression, and reverb – you'll also find the DeEsser and Pitch Correction plug-ins helpful in crafting that polished vocal performance your mix requires.

Logic's DeEsser is a handy way of taming any excessive sibilance (like the letters s and t) that might have been accentuated by a poor choice of microphone. The DeEsser works by analyzing the input and then applying a selective amount of gain reduction whenever problematic sibilance is heard. The Detector part of the interface is what should be used to spectrally locate the sibilance in the vocal. Set the Monitor mode to Det (detection) to tune into the specific frequency your signer's sibilance is occurring (usually somewhere in the region of 6 kHz). Now, flick to the Sens (sensitivity) monitor setting and adjust the sensitivity so that it only flicks on when the sibilance occurs. Moving the monitor to its OFF position, adjust the Suppressor parameter to the same parameter to that of the Detector and increase the strength to get the required amount of sibilance reduction.

(Continued)

The Pitch Correction plug-in is loosely based on the infamous Antares AutoTune plug-in and provides an on-the-fly means of correcting intonation problems in a vocal – or indeed, any other monophonic performance. At extremes, using the response setting on its 122 ms setting, the plug-in can be forced to produce the clichéd, quantized vocal effect so carelessly abused in the late 1990s! On softer sensitivity settings however, it can be a useful way of taming any problematic pitch drifts. To achieve the best results with the plug-in, you'll need to specify the key and scale the song is in. A number of presets for this are available, but you can also manually switch the notes in and out by clicking on the accompanying keyboard.

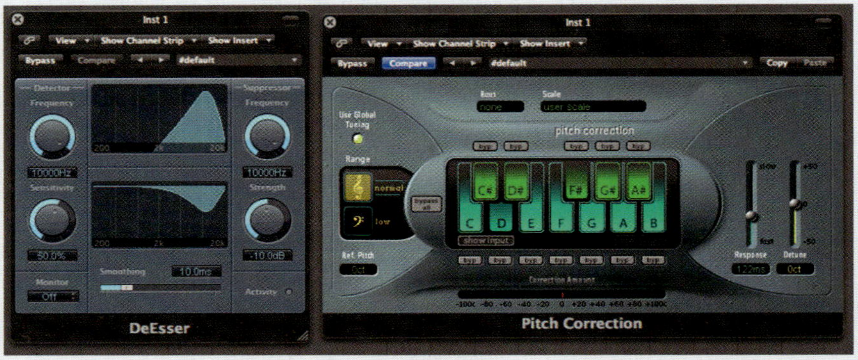

Figure 8.35 Logic's DeEsser and Pitch Correction plug-ins can aid a number of problems in relation to your vocals.

When two nodes are placed manually, you can also make use of the unique Automation Curve tool as a means of adding a degree or "curvature" to the line. Simply click and hold on the line between the nodes, using the Automation Curve tool, and drag above or below the line, or from side-to-side, to create one of four different adjustable curve shapes.

Where ranges of moves need to be moved or duplicated, you can also drag-enclose a number of nodes simply by holding Shift as you rubber band a group of nodes. With the nodes selected, they can then be moved en masse, or with the option key held down, duplicated to a new position. One other potential lifesaver, in case you've written in the automation moves but feel the need for them to be a couple of decibels higher or lower, is Logic's scaling feature. To scale a track's automation moves, use the Command key and drag up or down on the small bar graph meter as part of the track list. Look carefully at the selected values, and you should be able to spot the corresponding changes you've made.

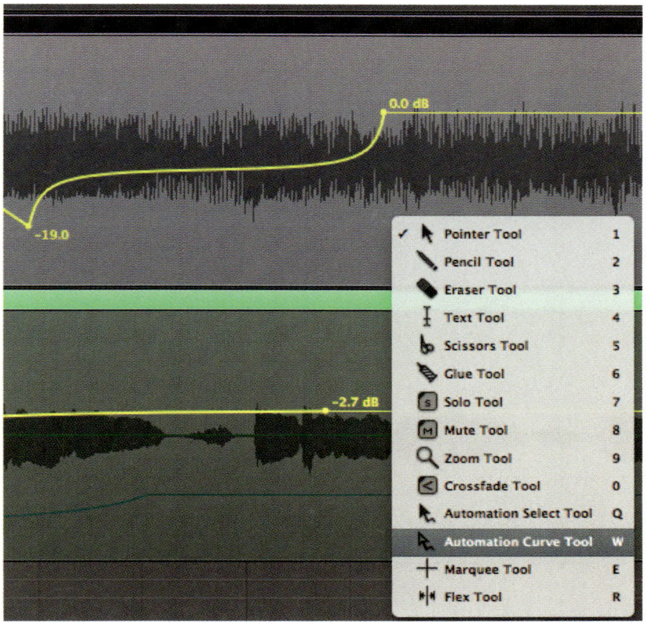

Figure 8.36 Use the Automation Curve tool to create neat and natural curves between automation modes.

Knowledgebase 3

Saving CPU: Freezing and More …

At some point in the mixing process, it's highly likely that you'll start to hit dreaded inevitability of a CPU overload. Dealing effectively with this problem could make the difference between a half-baked mix and a distinctive mix, using appropriate strategies to save CPU resources without compromising on sound quality. First, it helps if you're clear as to which plug-ins are the real CPU monsters in your session. For example, synthesizers like the ES2 and Sculpture can take up a big quota of resources, while the EXS24 (assuming it not making lots of use of its filter) is surprisingly processor efficient. In respect to plug-ins, Space Designer can be really CPU-hungry, especially if you're using longer IR files in excess of 2 s, while the Channel EQ is almost negligible on today's Intel-powered machines.

Arguably the first technique, therefore, is to look for any optimizations you can apply to the session that might improve overall CPU efficiency. For example, try turning off unused oscillators in the ES2, or removing excessive Space Designer use on inserts, in preference for a few instances on aux sends.

(Continued)

The next step is to use the track freeze option. Freezing effectively creates an audio bounce of the track in question, complete with all its inherent plug-in settings, and then deactivates the plug-ins accordingly. If you then try to edit any part of the track – either opening an instrument or plug-in or repositioning a region – Logic will remind you of its "frozen" status. If you do need to carry out the edits, simply unfreeze the track, make the modifications, and then refreeze.

Freezing itself is carried out with the small freeze icon as part of the track header. If you can't see the icon, select View > Configure Track Header and then add the freeze button into the set of options. Note that the freeze occurs across the full duration of the project, so it's worth moving back the project end marker to the real finish point of your track.

As an alternative, you can also carry out an audio bounce yourself, simply rendering complex virtual instruments as audio regions, saving off the instrument settings, and packing away the MIDI data (just in case you need to go back). This can be an effective solution earlier in the project where you intend to do lots of structural rearrangement with the regions at a later point.

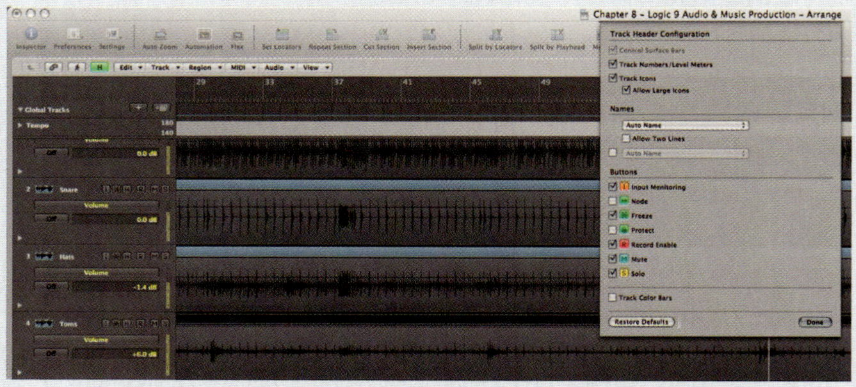

Figure 8.37 Use the Freeze function to render an off-line version of the track and so release valuable CPU resources.

8.14 The Automation Menu Options

Besides the various graphic tools for editing automation data, there's also an accompanying automation menu (Track > Track Automation) containing various options to delete moves, as well as some intriguing options to move data back

Figure 8.38 The Track Automation menu includes several powerful features, including the ability to move automation between its native track-based form and the region-based solution of Hyper Draw.

and forth between track-based automation and region-based automation (as previously discussed). Although it is, of course, easy enough to delete selective parts of automation using the eraser tool, the menu options make quick-and-easy to create "blank-slate" on individual automation lines, tracks, or indeed the whole song itself. Certainly, in situations where you might inherit a previous song file for a new project, this can be a great way of clearing out problematic automation moves.

Moving track-based automation to region-based automation is useful in situations where you'd like the automation permanently attached to a part, without having to continually specify for the automation data to be moved every time you realign or duplicate the part. Possibly the best "real-world" application of this would have to be the use of automation to control filter movements – maybe you've written a couple of distinctive filter movements that are as much part of the musicality in the region as the notes contained within the MIDI sequence. By switching the automation over to region-based (Track > Track Automation > Move Visible Track Automation Data To Region), you effectively lock the movements into the part, allowing yourself to forget about whether the data are moved or not. Of course, at any point, this same data can be brought back to track-based automation using Track > Track Automation > Move Visible Region Data to Track Automation.

Knowledgebase 4

Compression Circuit Types

The qualities of the different compressors are something that tends to excite many professional engineers, mainly as the different approaches to circuit design can achieve some radical differences in the types of compression achieved. Ultimately, this means that certain compressors tend to suit themselves to specific applications, bringing plenty of "character," as well as gain control, to the input they're processing. In the example of Logic's compressor plug-in, therefore, we're provided with Platinum, ClassA_R, ClassA_U, VCA, FET, and Opto.

Although the exact models from which these circuit types are derived aren't supplied, it is worth noting some of the key differences between these different options and how best to apply them. Opto represents the oldest compressor design, based on early models that used optical cells as part of their gain control circuitry. This unique design resulted in a degree of latency – both with respect to the Attack and Release on the compressor – that tends to deliver a more "musical" compression, as apposed to a harder, more aggressive gain control. The Opto circuit type, therefore, works well with bass sounds (that don't suit fast Attack and Release times), vocals, or anything that you want to retain a degree of musicality and lightness with.

Figure 8.39 The different circuit types approximate the unique sonic behavior of many classic types of compressor like the Urei 1176 or LA-2A.

The FET design, on the other hand, used Field Effect Transistors to create a more heavy-handed compression, particularly good at catching transients. The FET circuit model therefore, and to our ears the ClassA_U, can produce some really effective results on drums, especially when used across overheads. In this application, don't be afraid to use low threshold setting with "pumping" Attack and Release times for an aggressive, almost low-fi compression sound.

Although this provides a theoretical background to select the circuit type, the best approach is to use your ears. Try configuring some basic compression settings and then flick between the different compressor models to hear the marked differences in how they sound.

Walkthrough

Adding Compression

Step 1:
Insert a compressor across the instrument you want to process and start by establishing some basic settings. Working from the default positions, try finding a ratio and threshold setting that works for the instrument you're

Figure 8.40

(Continued)

trying to process. For example, for a gentle compression, use a medium threshold with soft ratio (1.5:1 through to 2:1) yielding about 2–3dB of gain reduction (another term for compression). For a harder compression effect, consider bringing up the ratio (4:1 or more) and lowering the threshold to achieve 6dB or more of gain reduction.

Step 2:
With these basic settings established, you can now start to refine the compression a little. Try adapting the Attack and Release settings to best suit the style of compression you want and the sound you're trying to squash. Slower settings (Attack 40ms, Release 400ms) tend to create a more natural progression in and out of gain reduction, although you might find the occasional loud transient slipping through the net. Faster Attack and Release settings (Attack 0–10ms, Release 100ms) produce a "pumping" effect, which tends to work well where you want the compression to sound more noticeable.

Figure 8.41

Step 3:
As compression leads to an overall loss in level, you need to raise the output to restore the overall signal to its original peak level (although, of course, the signal will be more compressed). Try bypassing the compressor, noting the meter readings, and then using the Gain parameter (with the

compressor active again) to restore the original level. Listen carefully to the compressed sound in the mix – is there enough compression to sit the instrument correctly? Is the compression too obvious? Further fine-tuning (maybe increasing the ratio or softening the Attack and Release) optimizes the compression for the instrument's position in the mix.

Figure 8.42

Logic Tips

Plug-In Delay Compensation

Any plug-in added into a channels signal path creates a small amount of delay or latency, through the extra processing required in producing the effect. Thankfully, however, Logic includes a feature to compensate for any delays incurred through plug-in processing, called plug-in delay compensation (PDC), available under the general tab of the audio preferences (Preferences > Audio). For users using standard audio unit plug-ins (either Logic's own or from other developers), stick to the "audio tracks and instruments setting." However, if you are running a processing-accelerator system like Universal Audio's UAD-1 or TC Electronics' PowerCore system, you may notice delays building up when you start to use buses to apply UAD-1

(Continued)

or PowerCore plug-ins. In these situations, change the preference to its All setting – Logic should now play the entirety of the session in time.

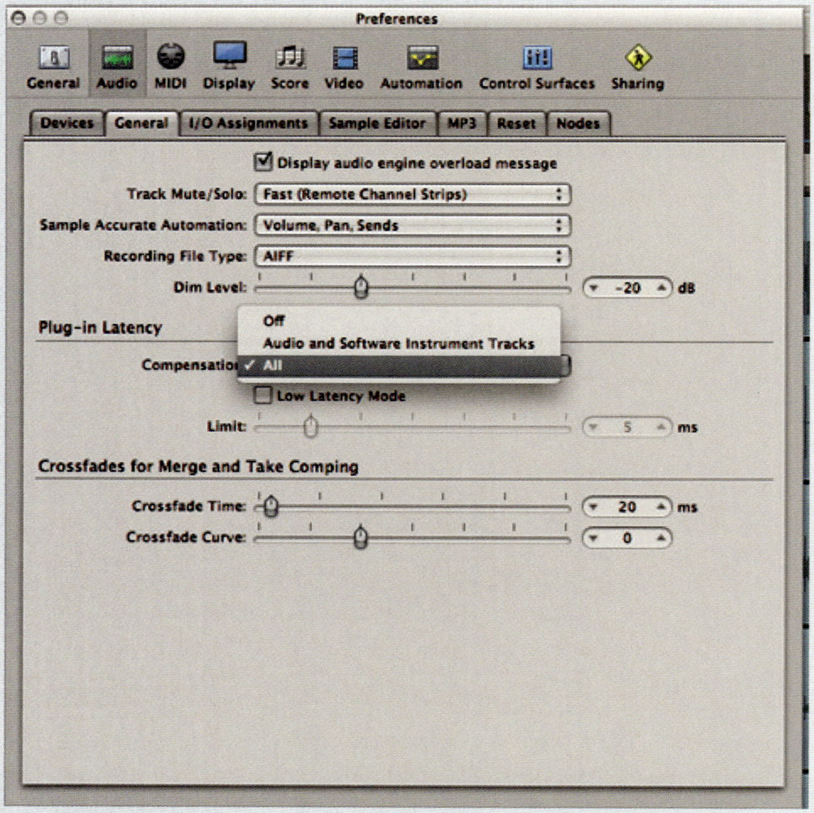

Figure 8.43 Changing the PDC settings is important for users of processing-accelerator systems such as Universal Audio's UAD or TC Electronics' PowerCore.

Plug-In Boxout 9

Pedalboard

Pedalboard is an ideal companion plug-in to Amp Designer and is designed to replicate the array of footpedals that guitarists might use to shape the sound of their guitar. As such, you'll probably want to place Pedalboard ahead of an instance of Amp Designer, although of course, you're free to use it anywhere along your signal path should you see fit. Indeed, the combination of both Pedalboard and Amp Designer shouldn't just be reserved for guitars, as the two plug-ins can work wonders on a range of sounds – from synth basses, to lo-fi breakbeats, and vocals!

Pedalboard is easy to use – simply drag the required pedals from the Pedal Browser on the right-hand side of the interface over to the pedal area found on the left-hand side of the plug-in. The signal path works from left to right, with each subsequent pedal adding the previous pedal's output. This means the order of the pedals can have a big effect on the eventual output, although it's generally best to place 'tone' pedals, like compressors, equalizers, and distortion pedals, earlier on in the signal path; with effects pedals, including flange, chorus, delay, and spring reverb later on in chain of footpedals.

As well as using a left-to-right 'serial' routing, Pedalboard also allows you to create parallel routing configurations where you can split the signal between two separate busses and then sum the results at the end of the chain of footpedals. For example, this makes it possible to layer two different types of distortion, rather than each distortion unit adding to the results of the previous pedal, or to send a split a sound to be processed by reverb and delay as separate entities.

You can assign a pedal to Bus B using the Routing area above Pedalboard – simply click on the pedal you want to move, and it switches to the upper Bus. Note a Mixer object now appears at the end of the signal path, allowing you to blend and pan the two discrete signal paths (A and B). Note that you can also manually insert a Splitter pedal to achieve similar result, although the added benefit here is that the Splitter pedal can also divide the sound by frequency, allowing each bus to process a different part of the frequency spectrum.

Figure 8.44 Pedalboard replicates the array of footpedals guitarists might use ahead of their amp to shape the basic sound of the guitar.